Disaster Resilience

A NATIONAL IMPERATIVE

Committee on Increasing National Resilience to Hazards and Disasters

Committee on Science, Engineering, and Public Policy

THE NATIONAL ACADEMIES

THE NATIONAL ACADEMIES PRESS
Washington, D.C.
www.nap.edu

THE NATIONAL ACADEMIES PRESS 500 Fifth Street, NW Washington, DC 20001

NOTICE: The project that is the subject of this report was approved by the Governing Board of the National Research Council, whose members are drawn from the councils of the National Academy of Sciences, the National Academy of Engineering, and the Institute of Medicine. The members of the committee responsible for the report were chosen for their special competences and with regard for appropriate balance.

This study was supported by the U.S. Army Corps of Engineers under award number W912HQ-10-C-0071, U.S. Department of Agriculture Forest Service under award number 09-DG-11221637'351, U.S. Department of Energy under award number DE-PI0000010, U.S. Department of Commerce National Oceanic and Atmospheric Administration under award number DG-133R-08CQ0062, Department of Homeland Security HSHQDC-10-C-00087, Federal Emergency Management Agency under award number HSFEHQ-11-C-1642, Department of the Interior U.S. Geological Survey under award number G104P00079, National Aeronautics and Space Administration under award number NNXIOAN3IG, and Community and Regional Resilience Institute and Oak Ridge National Laboratory under award number 4000090613. Any opinions, findings, conclusions, or recommendations expressed in this publication are those of the author(s) and do not necessarily reflect the views of the organizations or agencies that provided support for the project.

ISBN-13: 978-0-309-26150-0
ISBN-10: 0-309-26150-3
Library of Congress Control Number: 2012953228

Additional copies of this report are available for sale from the National Academies Press, 500 Fifth Street, NW, Keck 360, Washington, DC 20001; (800) 624-6242 or (202) 334-3313; http://www.nap.edu/ .

Cover: Conceptual design by Eric Edkin; design layout by Anne Rogers

Printed in the United States of America

THE NATIONAL ACADEMIES

Advisers to the Nation on Science, Engineering, and Medicine

The **National Academy of Sciences** is a private, nonprofit, self-perpetuating society of distinguished scholars engaged in scientific and engineering research, dedicated to the furtherance of science and technology and to their use for the general welfare. Upon the authority of the charter granted to it by the Congress in 1863, the Academy has a mandate that requires it to advise the federal government on scientific and technical matters. Dr. Ralph J. Cicerone is president of the National Academy of Sciences.

The **National Academy of Engineering** was established in 1964, under the charter of the National Academy of Sciences, as a parallel organization of outstanding engineers. It is autonomous in its administration and in the selection of its members, sharing with the National Academy of Sciences the responsibility for advising the federal government. The National Academy of Engineering also sponsors engineering programs aimed at meeting national needs, encourages education and research, and recognizes the superior achievements of engineers. Dr. Charles M. Vest is president of the National Academy of Engineering.

The **Institute of Medicine** was established in 1970 by the National Academy of Sciences to secure the services of eminent members of appropriate professions in the examination of policy matters pertaining to the health of the public. The Institute acts under the responsibility given to the National Academy of Sciences by its congressional charter to be an adviser to the federal government and, upon its own initiative, to identify issues of medical care, research, and education. Dr. Harvey V. Fineberg is president of the Institute of Medicine.

The **National Research Council** was organized by the National Academy of Sciences in 1916 to associate the broad community of science and technology with the Academy's purposes of furthering knowledge and advising the federal government. Functioning in accordance with general policies determined by the Academy, the Council has become the principal operating agency of both the National Academy of Sciences and the National Academy of Engineering in providing services to the government, the public, and the scientific and engineering communities. The Council is administered jointly by both Academies and the Institute of Medicine. Dr. Ralph J. Cicerone and Dr. Charles M. Vest are chair and vice chair, respectively, of the National Research Council.

www.national-academies.org

Committee on Increasing National Resilience to Hazards and Disasters

Susan L. Cutter (Chair), Carolina Distinguished Professor and Director, Hazards and Vulnerability Research Institute, University of South Carolina, Columbia

Maj. Gen. Joseph A. Ahearn (Retired), Senior Vice President, CH2M HILL Ltd, Colorado

Bernard Amadei, Professor of Civil Engineering, Department of Civil, Environmental and Architectural Engineering, University of Colorado at Boulder

Patrick Crawford, Director of Disaster Services, Feeding America, Chicago, Illinois

Gerald E. Galloway, Jr., Glenn L. Martin Institute Professor of Engineering, University of Maryland, College Park

Michael F. Goodchild, Professor, Department of Geography, University of California, Santa Barbara

Howard C. Kunreuther, James G. Dinan Professor of Decision Sciences & Public Policy, Wharton School, University of Pennsylvania, Philadelphia

Meredith Li-Vollmer, Risk Communication Specialist, Public Health Seattle and King County, Washington

Monica Schoch-Spana, Senior Associate, University of Pittsburgh Medical Center, Baltimore, Maryland

Susan C. Scrimshaw, President, The Sage Colleges, Troy, New York

Ellis M. Stanley, Sr., Director of Western Emergency Management Services, Dewberry LLC, Atlanta, Georgia

Gene Whitney, Energy Research Manager, Congressional Research Service, Washington, DC

Mary Lou Zoback, Consulting Professor, Stanford University, Stanford, California

Staff

Lauren Alexander-Augustine, Associate Executive Director, Division on Earth and Life Studies, and Director, Disasters Roundtable

Elizabeth A. Eide, Director, Board on Earth Sciences and Resources, and Study Director

Neeraj P. Gorkhaly, Research Associate

Eric J. Edkin, Senior Program Assistant

John H. Brown, Program Associate

Committee on Science, Engineering, and Public Policy

Richard N. Zare (Chair), Marguerite Blake Wilbur Professor in Natural Science, Department of Chemistry, Stanford University, Stanford, California
Linda M. Abriola (ex-officio), Dean of Engineering, Tufts University
Claude R. Canizares, Vice President for Research and Associate Provost and Bruno Rossi Professor of Experimental Physics, Massachusetts Institute of Technology
Moses H. W. Chan, Evan Pugh Professor of Physics, Pennsylvania State University
Ralph J. Cicerone (ex-officio), President, National Academy of Sciences
Paul Citron, Vice President (Retired), Technology Policy and Academic Relations, Medtronic, Inc.
Ruth A. David, President and Chief Executive Officer, ANSER (Analytic Services, Inc.)
Harvey V. Fineberg (ex-officio), President, Institute of Medicine
C. Dan Mote, Jr. (ex-officio), President Emeritus and Glenn Martin Institute Professor of Engineering, University of Maryland
Percy A. Pierre, Vice President and Professor Emeritus, Michigan State University
E. Albert Reece, Vice President for Medical Affairs, Bowers Distinguished Professor and Dean, School of Medicine, University of Maryland
Susan C. Scrimshaw, President, The Sage Colleges
William J. Spencer, Chairman Emeritus, SEMATECH
Michael S. Turner, Rauner Distinguished Service Professor, Kavli Institute for Cosmological Physics, The University of Chicago
Charles M. Vest (ex-officio), President, National Academy of Engineering
Nancy S. Wexler, Higgins Professor of Neuropsychology, Columbia University

Staff
Kevin Finneran, Director
Tom Arrison, Program Officer
Neeraj P. Gorkhaly, Research Associate
Marion Ramsey, Administrative Associate

Preface

Disaster resilience is everyone's business and is a shared responsibility among citizens, the private sector, and government. Increasing resilience to disasters requires bold decisions and actions that may pit short-term interests against longer-term goals. As a nation we have two choices. We can maintain the status quo and move along as we have for decades—addressing important, immediate issues such as the solvency of the National Flood Insurance Program, the most effective ways to discourage development in high-risk areas, and how to improve the speed and effectiveness disaster response. Or, we can embark on a new path—one that also recognizes and rewards the values of resilience to the individual, household, community, and nation. Such a path requires a commitment to a new vision that includes shared responsibility for resilience and one that puts resilience in the forefront of many of our public policies that have both direct and indirect effects on enhancing resilience.

The nation needs to build the capacity to become resilient, and we need to do this now. Such capacity building starts with individuals taking responsibility for their actions and moves to entire communities working in conjunction with local, state, and federal officials, all of whom need to assume specific responsibilities for building the national quilt of resilience. In the context of this report, the committee has used the term "community" in a very broad sense, encompassing the full range of potential communities—including local neighborhoods, family units, cities, counties, regions, or other entities. Defining a community as part of the nation's sense of collective resilience is a very site-specific endeavor, and the committee wanted to address this report toward the many kinds of communities that exist across the country.

Enhancing the nation's resilience to hazards and disasters is a laudable aspiration, but as is the case with such lofty goals, the devil is in the details. Although few would argue about the need to enhance the resilience of the nation and its communities to natural hazards, conflicts arise over how to move toward enhancing resilience, how to manage the costs of doing so, and how to assess its effectiveness. As we have seen, the costs of disasters are increasing as a function of more people and structures in harm's way as well as the effects of the extreme events themselves. These costs are being incurred at a time when more and more communities are financially constrained and unable to pay for essential services such as public safety and education. The choices that local communities have to make are thus difficult and not without some pain. At the same time, federal, state, and local governments have their own sets of constraints in terms of budget priorities, national interests, aging and declining infrastructure, and the political realities of implementing the kinds of changes needed to increase resilience. Disaster resilience may not be on the forefront of a political or institutional agenda until a disaster strikes one's own community. Political will and strong leadership are therefore essential to build resilience at any level.

The full range of roles and responsibilities, the broad stakeholder constituency, and even the iterative nature of building resilience are reflected in the sponsorship for this study, in the committee composition (Appendix A), and the information-gathering process used during this study. The nine study sponsors play different roles in monitoring and research, provision of data, community leadership, emergency management, disaster response, and short-term recovery. The committee comprises individuals with expertise in physical science and engineering, geographical science, social and behavioral science, economics, and public health, with professional experience from research, public policy, emergency and disaster management, nongovernmental organizations, the private sector, and government service. In many ways, resilience emerges as a topic that unites different groups with the goals of creating a common dialogue, reducing losses, and decreasing vulnerability to hazards and disasters. The committee and sponsors reflect this unity of purpose.

For this study, "national" does not equate to "federal." The stakeholders and audience for this study extend beyond the Washington, D.C. governmental community, and the experiential information necessary to understand national resilience lies in communities across the United States. To try to collect some of these regional experiences and information and the diversity of hazards faced in various parts of the country, the committee held three open meetings in New Orleans and the Mississippi Gulf Coast; Cedar Rapids and Iowa City, Iowa; and Southern California (Appendix B). Although many of the examples in the report are drawn from these three regions, the ideas and lessons are applicable to many communities across the nation. Discussions in workshops held in each of these three regions were supplemented by field excursions in the local communities to collect vital information about the successes and challenges people and institutions face in their efforts to become resilient to disasters. These three regions of the country were selected by the committee because each possesses a large amount of direct experience in building resilience through disaster preparedness, absorbing and responding to disasters, and in disaster recovery, adaptation, and mitigation.

Although the committee discussed very specific issues and broad hazards and disaster policies, we made a decision to offer recommendations that we, as a committee, felt were actionable by local, state, and federal interests and stakeholders in the short, medium, and long term. Implementation of these recommendations requires bipartisan support and involvement by private interests as well as those in the nonprofit sector.

Enhancing the nation's resilience will not be easy, nor will it be cheap. But the urgency is there and we need to begin the process now in order to build a national ethos that will make the nation safer, stronger, more secure, and more sustainable for our children and grandchildren.

Susan L. Cutter, Chair
July 2011

Acknowledgments

In addition to its own expertise, the committee relied on input from numerous external professionals and members of the public with extensive experience in public policy, emergency and disaster management, nongovernmental organizations, the private sector, government service, research, and personal and institutional responses to hazards and disaster events before, during, and after they occurred. These contributors provided data, references, and perspectives that assisted the committee in understanding the scope of the very broad issue of disaster resilience and the impact of decisions and actions that can increase or degrade the resilience of communities facing a variety of hazards and disasters. These individuals were very frank and open in providing important information to the committee without which it would have been impossible to develop this report. These individuals gave the committee distinct insights about what is happening at the local, state, and regional levels in terms of increasing disaster resilience.

We gratefully acknowledge these individuals and organizations and note that their thorough and helpful responses are brought forward throughout the report. The study's sponsors, the U.S. Army Corps of Engineers, U.S. Department of Agriculture Forest Service, U.S. Department of Energy, U.S. Department of Commerce National Oceanic and Atmospheric Administration, Department of Homeland Security and Federal Emergency Management Agency, Department of the Interior U.S. Geological Survey, National Aeronautics and Space Administration, and the Oak Ridge National Laboratory/Community and Regional Resilience Institute were particularly supportive and patient as the committee worked through this very challenging problem.

In addition, the committee would like to thank the following individuals who contributed to the study in different and meaningful ways:

In connection with the committee's Gulf Coast meeting, we thank Charles Allen III, Knox Andress, Justin Augustine, John Barry, Steven Bingler, Tap Bui, Garcia Bodley, Paul Byers, Commissioner Mike Chaney, Craig Colten, Maria Elisa Mandarim de Lacerda, Joseph Donchess, Mayor Garcia and Fire Chief Smith of Waveland, Mississippi, Greg Grillo, Kimberley Hoppe, Bill Howell, Natalie Jayroe, Pam Jenkins, Bob Klemme, Mary Claire Landry, Shirley Laska, Doug Meffert, Stephen Murphy, Earthea Nance, Eric Nelson, Tracy Nelson, May Nguyen, Allison Plyer, Julie Rochman, Ommeed Sathe, Ronald Schumann III, Tracie Sempier, Bill Stallworth, Marcia St. Martin,

Jonathan Thompson, and Frank Wise; community members of Village de L'Est and the owner of the café in which we held our discussion in East New Orleans; the Knight Nonprofit Center including Alice Graham, John Hosey, John Kelly, Rupert Lacy, Tom Lansford, Reilly Morse, Kimberly Nastasi, and Lori West.

In connection with the committee's meeting in Cedar Rapids and Iowa City, Iowa, we thank Jerry Anthony, Nancy Beers, Dee Brown, Christine Butterfield, Clark Christensen, Amy Costliow, Luciana Cunha, Lt. General Ron Dardis, Steve Dummeruth, Dave Elgin, Mark English, Kamyar Enshayan, Mitch Finn, Bill Gardam, Greg Graham, Donna Harvey, Benjamin Hoover, Patty Judge, Cindy Kaestner, Witold Krajewski, Carmen Langel, Kevin Leicht, Adam Lindenlaub, Alan Macvey, Liz Mathis, Jeff McClaran, Dave Miller, Tom Moore, Cornelia Mutel, Laura Myers, Doug Neumann, Corinne Peek-Asa, Lisa Pritchard, Marizen Ramirez, John Beldon Scott, Drew Skogman, Kyle Skogman, Megan Snitkey, Kathleen Stewart, Peter Thorne, James Throgmorton, Achilleas Tsakiris, Clint Twedt-Ball, Courtney Twedt-Ball, Terry Vaughan, Chad Ware, Larry Weber, Michael Wichman, Chuck Wieneke, Emily White, Leslie Wright, and Rick Wulfekuhle.

In connection with the committee's meeting in Irvine, California, we thank Mariana Amatullo, David Eisenman, Baruch Fischhoff, Alan Glennon, Mark Hansen, John Holmes, Lucy Jones, Sarah Karlinsky, Richard Little, Mike Morel, Javier Moreno, Leysia Palen, Chris Poland, Ezra Rapport, Roxanne Silver, Nalini Venkatasubramanian, and Matt Zook.

The helpful assistance we received with regard to planning and executing the field trips for the committee's regional meetings was also critical. We recognize the contributions from Senator Mary Landrieu who shared her welcoming remarks to open our workshop in the Gulf Coast. We also recognize and thank the city of New Orleans and Mayor Mitch Landrieu's office; the city of Cedar Rapids and Mayor Ron Corbett and City Manager Jeff Pomeranz; Cedar Rapids' Community Development Department including Christine Butterfield and Adam Lindenlaub; Leslie Wright from the United Way of East Central Iowa; Larry Weber from the University of Iowa; and John Holmes and the Port of Los Angeles. Their excellent cooperation and efforts to provide access to necessary information and localities greatly informed the committee's work.

At other stages of the study we also received very helpful contributions from Paul Brenner, Ben Billings, Laurie Johnson, Dennis Mileti, and Claire Rubin.

This report has been reviewed in draft form by individuals chosen for their diverse perspectives and technical expertise, in accordance with procedures approved by the National Research Council's (NRC) Report Review Committee. The purpose of this independent review is to provide candid and critical comments that will assist the institution in making its published report as sound as possible and to ensure that the report meets institutional standards for objectivity, evidence, and responsiveness to the study charge. The review comments and draft manuscript remain confidential to protect the integrity of the deliberative process.

We wish to thank the following individuals for their review of this report: Jacobo Bielak, Carnegie Mellon University; Christine Butterfield, City of Cedar Rapids-Iowa; Susan Curry, University of Iowa; Joseph Donovan, Beacon Hill Partners; Christopher Field, Carnegie Institution of Washington; Brian Flynn, Uniformed Services University of Health Sciences; Stephen Flynn, Northeastern University; Sandro Galea, Columbia University; Edward George, Massachusetts General Hospital; Jack Harrald, Virginia Polytechnic University; Bryan Koon, Florida Division of Emergency Management; John Krueger, Cherokee Nation Health Service; Burrell Montz, East Carolina University; Christopher Poland, Degenkolb Engineers; Barbara Reynolds, Centers for Disease Control; and Adam Rose, University of Southern California.

Although the reviewers listed above provided many constructive comments and suggestions, they were not asked to endorse the conclusions or recommendations nor did they see the final draft of the report before its release. The review of this report was overseen by Dr. Susan Hanson, Clark University (emeriti), and Dr. Mary Clutter, National Science Foundation (retired). Appointed by the NRC, they were responsible for making certain that an independent examination of this report was carried out in accordance with institutional procedures and that all review comments were carefully considered. Responsibility for the final content of this report rests entirely with the authoring committee and the institution.

CONTENTS

SUMMARY **1**

1. THE NATION'S AGENDA FOR DISASTER RESILIENCE **11**
Resilience: Why Now? 11
The National Imperative to Increase Resilience 13
Resilience Defined and the Role of this Study 16
On the Nation's Resilience Agenda 22
References 23

2 THE FOUNDATION FOR BUILDING A RESILIENT NATION: UNDERSTANDING, MANAGING, AND REDUCING DISASTER RISKS **25**
Understanding Risk 26
Managing Risk 28
Decision Making Under Risk and Uncertainty 38
Risk Management Strategies and Measures 43
Improving Resilience Through Risk Management 58
Knowledge and Data Needs 59
Summary and Recommendations 61
References 63

3 MAKING THE CASE FOR RESILIENCE INVESTMENTS: THE SCOPE OF THE CHALLENGE **67**
Introduction 67
Challenge of Resilience Decision Making for Community Leaders 69
The Scale and Scope of Disasters and Disaster Losses—An Urgent Problem 71
Knowledge and Data Needs 85
Summary and Recommendation 86
References 88

4 MEASURING PROGRESS TOWARD RESILIENCE **91**
The Need for Metrics and Indicators 91
Measures of U.S. National Resilience 94
International Efforts to Measure Resilience 104
The Committee's Perspective 110
Knowledge and Data Needs 111
Summary and Recommendation: Implementing a Measurement System 112
References 114

5	**BUILDING LOCAL CAPACITY AND ACCELERATING PROGRESS: RESILIENCE FROM THE BOTTOM UP**	**117**
	Whole Community Engagement	118
	Linking Private and Public Infrastructure Interests	127
	Communication to Build Resilience	134
	Zoning and Building Codes and Standards	144
	Research and Information Needs	150
	Summary and Recommendations	150
	References	153
6	**THE LANDSCAPE OF RESILIENCE POLICY: RESILIENCE FROM THETOP DOWN**	**159**
	Introduction	159
	Existing Federal Policies That Strengthen Resilience	160
	State and Local Authorities and Policies	182
	Unintended Consequences: Policies and Practices That Negatively Impact Resilience	186
	Resilience Policy Gaps and Needs	192
	Summary, Findings, and Recommendation	194
	References	195
7	**PUTTING THE PIECES TOGETHER: LINKING COMMUNITIES AND GOVERNANCE TO GUIDE NATIONAL RESILIENCE**	**197**
	Steps for Implementation	206
	References	208
8	**BUILDING A MORE RESILIENT NATION: THE PATH FORWARD**	**209**

APPENDIXES

A.	Committee and Staff Biographical Information	219
B.	Committee Meetings and Workshop Agendas	229
C.	Essential Hazard Monitoring Networks	241

Summary

No person or place is immune from disasters or disaster-related losses. Infectious disease outbreaks, acts of terrorism, social unrest, or financial disasters in addition to natural hazards can all lead to large-scale consequences for the nation and its communities. Communities and the nation thus face difficult fiscal, social, cultural, and environmental choices about the best ways to ensure basic security and quality of life against hazards, deliberate attacks, and disasters. Beyond the unquantifiable costs of injury and loss of life from disasters, statistics for 2011 alone indicate economic damages from natural disasters in the United States exceeded $55 billion, with 14 events costing more than a billion dollars in damages each.

One way to reduce the impacts of disasters on the nation and its communities is to invest in enhancing resilience. As defined in this report, resilience is *the ability to prepare and plan for, absorb, recover from, and more successfully adapt to adverse events.* Enhanced resilience allows better anticipation of disasters and better planning to reduce disaster losses—rather than waiting for an event to occur and paying for it afterward.

However, building the culture and practice of disaster resilience is not simple or inexpensive. Decisions about how and when to invest in increasing resilience involve short- and long-term planning and investments of time and resources prior to an event. Although the resilience of individuals and communities may be readily recognized after a disaster, resilience is currently rarely acknowledged *before* a disaster takes place, making the "payoff" for resilience investments challenging for individuals, communities, the private sector, and all levels of government to demonstrate.

The challenge of increasing national resilience has been recognized by the federal government, including eight federal agencies and one community resilience group affiliated with a National Laboratory who asked the National Research Council (NRC) to address the broad issue of increasing the nation's resilience to disasters. These agencies asked the NRC study committee to (1) define "national resilience" and frame the main issues related to increasing resilience in the United States; (2) provide goals, baseline conditions, or performance metrics for national resilience; (3) describe the state of knowledge about resilience to hazards and disasters; and (4) outline additional information, data, gaps, and/or obstacles that need to be addressed to increase the nation's resilience to disasters. The committee was also asked for recommendations about the necessary approaches to elevate national resilience to disasters in the United States.

This report confronts the topic of how to increase the nation's resilience to disasters through a vision of the characteristics of a resilient nation in the year 2030. The characteristics describe a more resilient nation in which

- Every individual and community in the nation has access to the risk and vulnerability information they need to make their communities more resilient.
- All levels of government, communities, and the private sector have designed resilience strategies and operation plans based on this information.
- Proactive investments and policy decisions have reduced loss of lives, costs, and socioeconomic impacts of future disasters.
- Community coalitions are widely organized, recognized, and supported to provide essential services before and after disasters occur.
- Recovery after disasters is rapid and the per capita federal cost of responding to disasters has been declining for a decade.
- Nationwide, the public is universally safer, healthier, and better educated.

The alternative, the status quo, in which the nation's approaches to increasing disaster resilience remain unchanged, is a future in which disasters will continue to be very costly in terms of injury; loss of lives, homes, and jobs; business interruption; and other damages.

Building resilience toward the 2030 future vision requires a paradigm shift and a new national "culture of disaster resilience" that includes components of

(1) Taking responsibility for disaster risk;
(2) Addressing the challenge of establishing the core value of resilience in communities, including the use of disaster loss data to foster long-term commitments to enhancing resilience;
(3) Developing and deploying tools or metrics for monitoring progress toward resilience;
(4) Building local, community capacity because decisions and the ultimate resilience of a community are driven from the bottom up;
(5) Understanding the landscape of government policies and practices to help communities increase resilience; and
(6) Identifying and communicating the roles and responsibilities of communities *and* all levels of government in building resilience.

A set of six actionable recommendations (see Box S-1 at the close of the Summary) are described that will help guide the nation toward increasing national resilience from the local community through to state and federal levels. The report has been informed by published information, the committee's own expertise, and importantly, by experiences shared by communities in New Orleans and the Mississippi Gulf Coast, Cedar Rapids and Iowa City, Iowa, and Southern California, where the committee held open meetings.

UNDERSTANDING, MANAGING, AND REDUCING DISASTER RISK

Understanding, managing, and reducing disaster risks provide a foundation for building resilience to disasters. Risk represents the potential for hazards to cause adverse effects on our life; health; economic well-being; social, environmental, and cultural assets; infrastructure; and the services expected from institutions and the environment. Risk management is a continuous process that identifies the hazard(s) facing a community, assesses the risk from these hazards, develops and implements risk strategies, reevaluates and reviews these strategies, and develops and adjusts risk policies. The choice of risk management strategies requires regular reevaluation in the context of new data and models on the hazards and risk facing a community, and changes in the socioeconomic and demographic characteristics of a community, as well as the community's goals. Although some residual risk will always be present, risk management strategies can help build capacity for communities to become more resilient to disasters.

A variety of tools exist to manage disaster risk including tangible structural (construction-related) measures such as levees and dams, disaster-resistant construction, and well-enforced building codes, and nonstructural (nonconstruction-related) measures such as natural defenses, insurance, zoning ordinances, and economic incentives. Structural and nonstructural measures are complementary and can be used in conjunction with one another. Importantly, some tools or actions that can reduce short-term risk can potentially increase long-term risk, requiring careful evaluation of the risk management strategies employed. Risk management is at its foundation a community decision, and the risk management approach will be effective only if community members commit to use the risk management tools and measures made available to them.

THE CHALLENGE OF MAKING INVESTMENTS IN RESILIENCE

Demonstrating that community investments in resilience will yield measurable short- and long-term benefits that balance or exceed the costs is critical for sustained commitment to increasing resilience. The total value of a community's assets—both the high-value structural assets and those with high social, cultural, and/or environmental value—call for a decision-making framework for disaster resilience that addresses both quantitative data and qualitative value assessments. Ownership of a community's assets is also important; ownership establishes the responsibility for an asset and, therefore, the need to make appropriate resilience investments to prepare and plan for hazards and risks. Presently, little guidance exists for communities to understand how to place meaningful value on all of their assets. Particularly during times of economic hardship, competing demand for many societally relevant resources (education, social services) can be a major barrier to making progress in building resilience in communities.

Accessing and understanding the historical spatial and temporal patterns of economic and human disaster losses in communities in the United States are ways for communities to understand the full extent of the impact of disasters and thereby motivate community efforts to increase resilience. Historical patterns of disaster losses provide some sense of the magnitude of the need to become more disaster resilient. The geographic patterns of disaster losses—e.g., human fatalities, property losses, and crop losses—illustrate where the impacts are the greatest, what challenges exist in responding to and recovering from disasters, and what factors drive exposure and vulnerability to hazards and disasters. Although existing loss databases in the United States are useful for certain kinds of analyses, improvement in measurements, accuracy, and consistency are needed. Furthermore, the nation lacks a national repository for all-hazard event and loss data, compromising the ability of communities to make informed decisions about where and how to prioritize their resilience investments.

MEASURING PROGRESS TOWARD RESILIENCE

Without some numerical means of assessing resilience it would be impossible to identify the priority needs for improvement, to monitor changes, to show that resilience had improved, or to compare the benefits of increasing resilience with the associated costs. The measurement of a concept such as resilience is difficult, requiring not only an agreed-upon metric, but also the data and algorithms needed to compute it. The very act of defining a resilience metric, and the discussions that ensue about its structure, helps a community to clarify and formalize what it means by the concept of resilience, thereby raising the quality of debate. The principles that resilience metrics can entail are illustrated by some existing national and international indicators or frameworks that address measurement of the resilience of different aspects of community systems. The Leadership in Energy and Environmental Design for developers, owners, and operators of buildings is one example. Comparison of the strengths and challenges of a variety of different frameworks for measuring resilience suggests that the critical dimensions of an encompassing and consistent resilience measurement system are

- Indicators of the ability of critical infrastructure to recover rapidly from impacts;

- Social factors that enhance or limit a community's ability to recover, including social capital, language, health, and socioeconomic status;
- Indicators of the ability of buildings and other structures to withstand earthquakes, floods, severe storms, and other disasters; and
- Factors that capture the special needs of individuals and groups, related to minority status, mobility, or health status.

Presently, the nation does not have a consistent basis for measuring resilience that includes all of these dimensions. Until a community experiences a disaster and has to respond to and recover from it, demonstrating the complexity, volume of issues, conflicts and lack of ownership are difficult. A national resilience scorecard, from which communities can then develop their own, tailored scorecards, will make it easier for communities to see the issues they will face prior to an event and can support necessary work in anticipation of an appropriate resilience-building strategy. A scorecard will also allow communities to ask the right questions in advance of a disaster.

BUILDING LOCAL CAPACITY AND ACCELERATING PROGRESS: RESILIENCE FROM THE BOTTOM UP

National resilience emerges, in large part, from the ability of local communities with support from all levels of government and the private sector to plan and prepare for, absorb, respond to, and recover from disasters and adapt to new conditions. Bottom-up interventions—the engagement of communities in increasing their resilience—are essential because local conditions vary greatly across the country; the nation's communities are unique in their history, geography, demography, culture, and infrastructure; and the risks faced by every community vary according to local hazards. Some universal steps can aid local communities in making progress to increase their resilience and include:

- Engaging the whole community in disaster policymaking and planning;
- Linking public and private infrastructure performance and interests to resilience goals;
- Improving public and private infrastructure and essential services (such as health and education);
- Communicating risks, connecting community networks, and promoting a culture of resilience;
- Organizing communities, neighborhoods, and families to prepare for disasters;
- Adopting sound land-use planning practices; and

- Adopting and enforcing building codes and standards appropriate to existing hazards.

Community coalitions of local leaders from public and private sectors, with ties to and support from federal and state governments, and with input from the local citizenry, become very important in this regard. Such coalitions can be charged to assess the community's exposure and vulnerability to risk, to educate and communicate risk, and to evaluate and expand the community's capacity to handle such risk. A truly robust coalition would have at its core a strong leadership and governance structure, and people with adequate time, skill, and dedication necessary for the development and maintenance of relationships among all partners in the community.

THE LANDSCAPE OF RESILIENCE POLICY: RESILIENCE FROM THE TOP DOWN

Strong governance at all levels is a key element of resilience and includes the making of consistent and complementary local, state, and federal policies. Although resilience at its core has to be carried forward by communities, communities do not exist under a single authority in the United States, and function instead under a mix of policies and practices implemented and enforced by different levels of government. Policies that make the nation more resilient are important in every aspect of American life and economy, and not just during times of stress or trauma. A key role of policies designed to improve national resilience is to take the long-term view of community resilience and to help avoid short-term expediencies that can diminish resilience.

Certain policies of the federal Executive Branch, including Presidential Directives and Executive Orders, policies initiated by federal agencies, and policies of the Legislative Branch can and do function to help strengthen resilience. Presidential Policy Directive-8 (PPD-8) calls upon the Department of Homeland Security to embrace systematic preparation against all types of threats, including catastrophic natural disasters. Because the scope of resilience is sometimes not fully appreciated, some who contemplate national resilience policy think first of the Stafford Act and its role in disaster response and recovery. Although the Stafford Act does provide guidance for certain responsibilities and actions in responding to a disaster incident, national resilience transcends the immediate impact and disaster response and therefore grows from a broader set of policies. Many of the critical policies and actions required for improved national resilience are also enacted and implemented at the state and local levels.

Policies at all levels of governance do exist to enhance resilience; however, some government policies and practices can also have unintended consequences that negatively affect resilience. Furthermore, gaps in policies and programs among federal agencies exist for all parts of the resilience process—including disaster preparedness, response, recovery, mitigation, and adaptation, as well as research, planning, and community assistance. Although some of these gaps are the result of the legislative authorization within which agencies are directed to operate, the roles and responsibilities for building resilience are not effectively coordinated by the federal government, either through a single agency or authority, or through a unified vision.

Community resilience is broad and complex, making it difficult to codify resilience in a single comprehensive law. Rather, infusing the principles of resilience into all the routine functions of the government at all levels and through a national vision is a more effective approach.

LINKING COMMUNITY AND GOVERNANCE TO GUIDE NATIONAL RESILIENCE

Increased resilience cannot be accomplished by simply adding a cosmetic layer of policy or practice to a vulnerable community. Long-term shifts in physical approaches (new technologies, methods, materials, and infrastructure systems) and cultural approaches (the people, management processes, institutional arrangements, and legislation) are needed to advance community resilience. Resilience to disasters rests on the premise that all aspects of a community—its physical infrastructure, its socioeconomic health, the health and education of its citizens, and its natural environment—are strong. This kind of systemic strength requires that the community members work in concert and in such a way that the interdependencies among them provide strength during a disaster event.

Communities and the governance network of which they are a part are complex and dynamic systems that develop and implement resilience-building policies through combined effort and responsibility. Experience in the disaster management community suggests that linked bottom-up and top-down networks are important for managing risk and increasing resilience. Key interactions within the nation's resilience "system" of communities and governance can be used to help identify specific kinds of policies that can increase resilience and the roles and responsibilities of the actors in government, the private sector, and communities for implementing these policies. For example, to understand hazards or threats and their processes, research and science and technology policies allow federal and state agencies to coordinate efforts on detection and monitoring activities that can be used by regional and local governing bodies, the private sector, and communities to evaluate and address their hazards and risks. Identifying resilience policy areas, identifying those in community and government responsible for coordinating activities in those areas, and identifying the recipients of the information or services resulting from those activities reveal strengths and gaps in the nation's resilience "system."

Advancing resilience is a long-term process, but can be coordinated around visible, short-term goals that allow individuals and organizations to measure or mark their progress toward becoming resilient and overcoming these gaps. However, as a necessary first step to strengthen the nation's resilience and provide the leadership to establish a national "culture of resilience," a full and clear commitment to disaster resilience by the federal government is essential.

BUILDING A MORE RESILIENT NATION: THE PATH FORWARD

No single sector or entity has ultimate responsibility for improving national resilience. No specific federal agency has all of the authority or responsibility, all of the appropriate skill sets, or adequate fiscal resources to address this growing challenge. An important responsibility for increasing national resilience lies with residents and their communities. Input, guidance, and commitment from all levels of government and from the private sector, academia, and community-based and nongovernmental organizations are needed throughout the entire process of building more resilient communities. The report frames six recommendations (Box S-1) that can help guide the nation in advancing collective, resilience-enhancing efforts in the coming decades.

BOX S-1
Summary Recommendations

***Recommendation 1*: Federal government agencies should incorporate national resilience as a guiding principle to inform the mission and actions of the federal government and the programs it supports at all levels.**

***Recommendation 2*: The public and private sectors in a community should work cooperatively to encourage commitment to and investment in a risk management strategy that includes complementary structural and nonstructural risk-reduction and risk-spreading measures or tools. Such tools might include an essential framework (codes, standards, and guidelines) that drives the critical structural functions of resilience and investment in risk-based pricing of insurance.**

***Recommendation 3:* A national resource of disaster-related data should be established that documents injuries, loss of life, property loss, and impacts on economic activity. Such a database will support efforts to develop more quantitative risk models and better understand structural and social vulnerability to disasters.**

Recommendation 4: **The Department of Homeland Security in conjunction with other federal agencies, state and local partners, and professional groups should develop a National Resilience Scorecard.**

Recommendation 5: **Federal, state, and local governments should support the creation and maintenance of broad-based community resilience coalitions at local and regional levels.**

Recommendation 6**: All federal agencies should ensure that they are promoting and coordinating national resilience in their programs and policies. A resilience policy review and self-assessment within agencies and strong communication among agencies are keys to achieving this kind of coordination.**

Increasing disaster resilience is an imperative that requires the collective will of the nation and its communities. Although disasters will continue to occur, actions that move the nation from reactive approaches to disasters to a proactive stance where communities actively engage in enhancing resilience will reduce many of the broad societal and economic burdens that disasters can cause.

"Anticipating and preparing for potential problems rather than reacting to events puts us in a better place when dealing with natural disasters."
Citizen from Curry County, Oregon

1

The Nation's Agenda for Disaster Resilience

RESILIENCE: WHY NOW?

In 2011 the United States was struck with multiple disasters including 14 weather- and climate-related events that each caused more than $1 billion in damages[1] (Figure 1.1). Statistics indicate that total economic damages from all natural disasters in 2011 exceeded $55 billion in property damage, breaking all records since these data were first reported in 1980 (NCDC, 2012). Cumulatively, nearly 600 Americans died[2] and many thousands of households were temporarily or permanently displaced by events that included blizzards, tornadoes, drought, flooding, hurricanes, and wildfires (Figure 1.2). Natural disasters have continued in 2012 as this report was being written—with tornadoes, massive wildfires, and flooding and wind damage affecting millions of people in the nation. These events have had local and national ramifications, and effects from them have been felt across large geographic areas and large portions of the population through injuries and death, destruction of homes and businesses, displacement of people, interruption of business , disruptions in transportation, job losses, and greater demands on federal and state resources. These disasters demonstrate very clearly the interconnectedness of natural and human systems and infrastructure and the strengths and frailties of these connections.

No person or place is immune from disasters or disaster-related losses. Infectious disease outbreaks, acts of terrorism, social unrest, or financial disasters in addition to natural hazards can all lead to large-scale consequences for the nation and its communities. Communities and the nation thus face difficult fiscal, social, cultural, and environmental choices about the best ways to ensure basic security and quality of life against hazards, deliberate attacks, and

[1] http://www.noaa.gov/extreme2011/
[2] http://www.emdat.be/result-country-profile?disgroup=natural&country=usa&period=2011$2011.

disasters. One way to reduce the impacts of disasters on the nation and its communities is to invest in enhancing resilience.

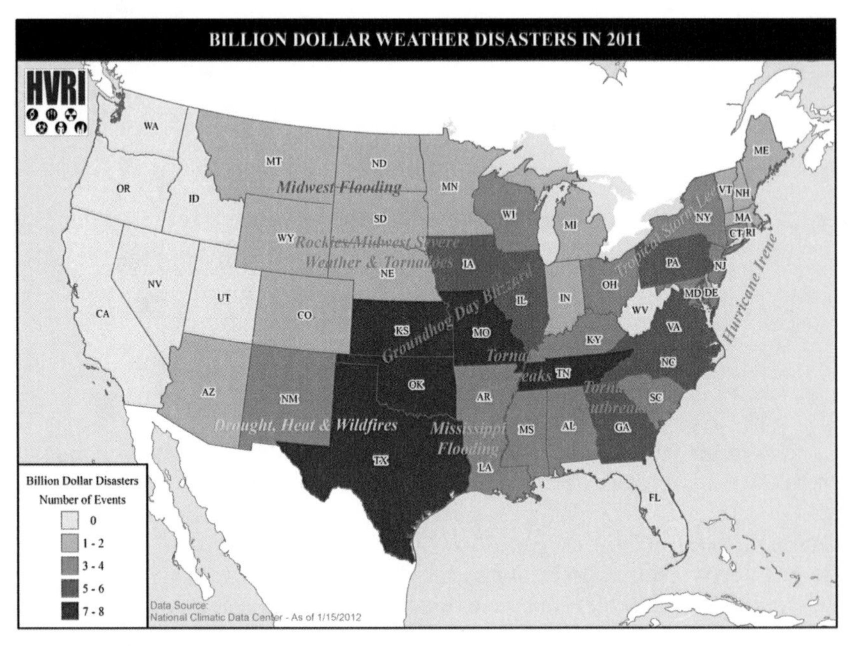

FIGURE 1.1 Areas in the United States affected by large weather disasters in 2011. HVRI = Hazards and Vulnerability Research Institute. Source: S. Cutter/HVRI.

The large amounts of money the federal government spends in responding to disasters are one indicator of the urgency of the need to increase the nation's resilience to these events. These expenditures are borne by the entire nation and have been growing steadily for the last 60 years, both in absolute terms and on a per capita basis. For example, in 1953, the first year of presidential disaster declarations, federal expenditures totaled $20.9 million (adjusted to 2009 dollars) or $0.13 per person. In 2009, with many more disaster declarations, the federal government conservatively spent $1.4 billion on federal disaster relief or the equivalent of about $4.75 per person.[3] The past two decades in particular show highly devastating and costly events to the nation's treasury: the 1994 Northridge earthquake led to federal expenditures of $11.6 billion in disaster relief, relief costs for the 2001 World Trade Center attack totaled $13.3 billion, and Hurricane Katrina alone in 2005 led to more than $48.7 billion in federal disaster relief costs. Importantly, these expenditures do not even include insured property or business interruption losses, which

[3] Computed from Federal Emergency Management Agency Presidential Disaster Declaration Data with totals adjusted to 2009 dollars.

otherwise significantly increase the total economic impact of these events. For example, property damage at the World Trade Center stemming from the 9/11 terrorist attacks amounted to $23 billion, but the costs for business interruption are estimated to have been around $100 billion (Rose and Blomberg, 2010).

FIGURE 1.2 Tornado damage in Joplin, Missouri, from the May 2011 tornado that struck the area, killing 159 people and injuring more than 1,000 others. The tornado was the single deadliest in U.S. history since such records have been kept. The tornado was 1 mile wide and traveled 22 miles on the ground (NOAA, 2011). Source: Charlie Riedel/AP Photo.

What happens to the magnitude of these losses of lives, livelihoods property, and community in the future as our population increases and our infrastructure ages and expands if we maintain the status quo and our nation does not improve its resilience to hazards and disasters? What does effective disaster resilience look like for the nation, for our communities, and for our families? What steps need to be taken to become more resilient in the near and long term?

THE NATIONAL IMPERATIVE TO INCREASE RESILIENCE

Decisions by communities, states, regions, and the nation regarding whether to invest in building resilience are difficult. If building the culture and practice of disaster resilience were simple and inexpensive, the nation would likely have taken steps to become more resilient already. Making the choice either to proceed with the status quo—where concerted investments and planning do not take place throughout the country to increase disaster resilience—or to make conscious decisions and investments to build more resilient communities is weighted by a few central points:

1) Disasters will continue to occur, whether natural or human-induced, in all parts of the country;
2) The population will continue to grow and age as will the number and size of communities; in some regions population decline and the number and size of communities will create a different set of challenges as tax bases decline;
3) Demographic data demonstrate that more people are moving to coastal and southern regions—areas with a high number of existing hazards such as droughts and hurricanes;
4) Public infrastructure is currently aging beyond acceptable design limits;
5) Infrastructure such as schools, public safety, and public health that are essential to communities are facing economically difficult times as the population grows and ages;
6) Economic and social systems are becoming increasingly interdependent and thus increasingly vulnerable should a key part of the system be disrupted;
7) Risk cannot be eliminated completely, so some residual risk will continue to exist and require management;
8) Impacts of climate change and degradation of natural defenses such as coastal wetlands make the nation more vulnerable.

This report suggests some of the characteristics of a *resilient nation* in the year 2030. This future vision of characteristics that the United States might have in 2030 requires alternative kinds of decisions and investments that will lead to a more resilient nation:

Characteristics of a Resilient Nation in the Year 2030

In 2030, the nation, from individuals to the highest levels of government, has embraced a "culture of resilience." Information on risks and vulnerability to individuals and communities is transparent and easily accessible to all. Proactive investments and policy decisions including those for preparedness, mitigation, response, and recovery have reduced the loss of lives, costs, and socioeconomic impacts of disasters. Community coalitions are widely organized, recognized, and supported to provide essential services before and after disasters occur. Recovery after disasters is rapid and includes funding from private capital. The per capita federal cost of responding to disasters has been declining for a decade.

The nation has an important stake in realizing the characteristics of this future vision. Achieving this kind of resilience in two decades encompasses actions and decisions at all levels of government, in the private sector, and in communities including, for example,

- Action by federal agencies to incorporate disaster risk and resilience in their policies and activities. Such actions could include strong support for monitoring activities and data collection for natural and human systems (Appendix C), and transfer and communication of resilience-related research to states, regions, and communities. Such data and research are critical for quantifying risk and measuring progress for resilience.
- Consistent federal assistance for community resilience based on loss avoidance or disaster risk reduction, rather than primarily on post-disaster relief.
- Nationwide infrastructure upgrades by both the private sector and all levels of government to meet 21st century technology needs and building codes and to encompass disaster-resilient designs.
- Increased nationwide increased support for public safety, health, and education as well as information systems.
- Commitment from local city and county officials to maintain and advocate land use, zoning plans, and construction codes that explicitly enhance resilience and emphasize working with the natural environment and valuing natural environmental defenses.
- Recognition at all levels of government and within communities that communities are part of a system that includes both the natural environment and other communities. Implicit in this recognition is the idea that actions to enhance a single community's resilience, such as constructing a flood defense system, may have positive and negative impacts on surrounding communities over time.
- Realization by individuals and communities that they provide their own first line of defense against disasters, including mutual assistance and governance structures designed to manage crises cooperatively.

This resilient future also includes understanding the economic benefits of resilience, such as the cost savings of mitigation, and valuing the protective functions and services of ecosystems. The costs for short-term mitigation alone can reduce much greater, longer-term losses. For example, the Multihazard Mitigation Council (2005) found that every dollar spent on pre-event mitigation related to earthquakes, wind, and flooding saved about $4 in post-event damages. Furthermore, the planning and preparation for one type of disaster (such as the nuclear accident planning experience in Cedar Rapids, Iowa—see Box 2.4), can reap benefits for other types of disasters or unexpected adverse events.

An alternative to the resilience vision is the current path of the nation—the status quo in which innovations are not made to increase the nation's resilience to hazards and disasters. Unless this current path in the nation's approach toward hazards and disasters is changed, data suggest that the cost of disasters will continue to rise both in absolute dollar amounts and in the losses to the social, cultural, and environmental systems that are part of each community. Communities that continue to build in areas such as floodplains, wetlands, and coastal zones may experience greater impacts from flooding, hurricanes, and sea-level rise (e.g., NRC, 2012). Continued reliance on structural or engineered solutions to control nature rather than to work in concert with natural systems may transfer and enhance disaster risks across geographic areas and through time. Businesses and households may remain vulnerable because of inadequate building and zoning codes. Vulnerable people such as the elderly or those with specific health issues will need more extensive and expensive assistance in a disaster. Increased property, job, and crop losses may result in greater demand for disaster relief funding from the federal government.

The various points relevant to increasing the nation's resilience, including the characteristics of a resilient nation in 2030, are developed in later chapters.

RESILIENCE DEFINED AND THE ROLE OF THIS STUDY

Many people have heard and used the term "resilience," perhaps in describing how an individual or a city or a nation showed great strength under adversity, or bounced back after some unexpected tragedy. After such events, an individual or city or nation can become stronger, its approaches and institutions more flexible, and its citizens and communities more capable of withstanding the next adverse circumstance. In addressing the broad topic of resilience, articulating what is meant by the term is important. This report defines resilience as *the ability to prepare and plan for, absorb, recover from, and more successfully adapt to adverse events* (Definition 1.1).

Definition 1.1

Resilience: The ability to prepare and plan for, absorb, recover from, or more successfully adapt to actual or potential adverse events.[4]

---

[4] This definition was developed by the study committee based on the extant literature and is consistent with the international disaster policy community (UNISDR, 2011), U.S. governmental agency definitions (SDR, 2005; DHS Risk Steering Committee, 2008; PPD-8, 2011), and NRC (2011).

This report considers resilience to disasters to encompass both natural and human-induced events. However, most of the data and information on disasters relate to those with natural causes and the report is weighted toward using those events as examples. The report addressed the importance of an "all-hazards" approach to resilience that encompasses the idea that planning for one kind of hazard or disaster event can increase the resilience of a community in the face of a different kind of event. When this concept of resilience is applied to hazards and disasters, whether natural or human-induced, being able to anticipate, withstand, and recover from such events with minimal human and economic losses becomes a very desirable goal. Further, being resilient to hazards and disasters is a condition toward which all communities and the nation can justifiably aspire. Despite efforts to reduce the impact of natural disasters, however, the United States as a whole is not fully resilient to disasters.

Communities and the nation can be better protected and strengthened by increasing resilience to disasters just as individuals take preventive measures to protect the human body against illness and disease (Box 1.1). A healthy body is not simply a composite of individual functioning systems. All the systems work together. In a similar way, the dynamic physical, social, political, economic, and environmental systems in resilient communities work and function together.

BOX 1.1
Why Effective Community Resilience Is Similar to a Healthy Human Body

Communities can be viewed as a set of interrelated systems that share a common vision, and the overall resilience of communities may be viewed in much the same way as the overall health of the human body. A human body relies on the integrated functioning of its shared systems—like the skeletal, nervous, and immune systems—to maintain health and resist disease and injury. Similarly, communities depend on a number of interrelated systems for economic stability and growth, commerce, education, communication, population wellness, energy, and transportation. The relative "health" of community systems will determine how well a community can withstand disruptive events. If a community has weakened infrastructure, like a human body with a compromised immune system, it will not withstand trauma as well as one in good health.

In both human health and community resilience, investments in maintaining health and building strength reduce the requirement for very expensive treatment and recovery. Health providers now know that prevention is a less expensive pathway than is treatment after the onset of an illness. In the same way, investment in community resilience before a disaster occurs may help a community reduce or avoid monumental recovery and restoration costs after the event has taken place.

When a community has been destroyed by disaster (unlike the human body that has died from disease) it is sometimes possible to bring it back to life. In all cases, though, avoiding destruction in the first place is cheaper, easier, and less traumatic over the long term than resuscitating a devastated community. Post-event mitigation, like remaining free from fatal illness, requires conscious, steady, and organized investment in resilience by those in charge of the care of a community and by the community itself.

This analogy can be extended to the idea that, just like a healthy body is better able to resist disease, a healthy community is better able to prepare for, absorb, and recover from a disaster. For example, infrastructure such as health care with broad access implies a population whose health problems are controlled and/or prevented to the extent possible. A robust health infrastructure enhances resilience, and provides data essential to the early detection of naturally occurring or terrorist-induced epidemics and environmental hazards.

Disaster resilience as an integrated part of community or government decision making is a relatively new concept that is only now being broadly or explicitly adopted through efforts such as Presidential Policy Directive-8 (PPD-8; see below and Chapter 6). Although many efforts have been made to understand disaster resilience and its benefits at various scales (Box 1.2), implementation of approaches to increase disaster resilience in communities has not been consistent nationwide.

The *process* of building disaster resilience requires continuous assessment, planning, and refinement by communities and all levels of government; resilience is not a task that can be marked as "completed." No

BOX 1.2
What Is Resilience?

Although resilience with respect to hazards and disasters has been part of the research literature for decades (White and Haas, 1975; Mileti, 1999), the term first gained currency among national governments in 2005 with the adoption of The Hyogo Framework for Action by 168 members of the United Nations to ensure that reducing risks to disasters and building resilience to disasters became priorities for governments and local communities (UNISDR, 2007). The literature has since grown with new definitions of resilience and the entities or systems to which resilience refers (e.g., ecological systems, infrastructure, individuals, economic systems, communities) (Bruneau et al., 2003; Flynn, 2007; Gunderson, 2009; Plodinec, 2009; Rose, 2009; Cutter et al., 2010). Disaster resilience has been described as a *process* (Norris et al., 2008; Sherrieb et al., 2010), an *outcome* (Kahan et al., 2009), or both (Cutter et al., 2008), and as a term that can embrace inputs from engineering and the physical, social, and economic sciences (Colten et al., 2008).

perfect end state or end condition of resilience exists. In fact, building resilience means building strong communities that contain adequate essential public and private services including schools, transportation, health care, utilities, roads and bridges, public safety, and businesses. A common understanding of what resilience means for a community, a set of achievable milestones and goals, the approaches for reaching those milestones, and agreed-upon measures of progress are thus required by the people, businesses, and government agencies associated with that community. Resilience also requires that people, businesses, and government agencies recognize their roles and responsibilities, individually and collectively, and act on these roles and responsibilities to help make their communities more resilient. While local and state institutions grapple with specific issues within their communities, for example, federal agencies provide the data, knowledge, tools, and assistance that are needed by all communities to help them become more resilient. A community's citizens and the private sector also have important roles and responsibilities to increase resilience. These roles and responsibilities, including the data and tools needed to increase resilience, are described in detail later in this report (see also Appendix C).

Enhancing the nation's resilience to disasters is a national imperative for the stability, progress, and well-being of the nation that can benefit the nation economically, environmentally, and from a national security perspective. However, the challenge of increasing national resilience is profound. These challenges were recognized collectively by eight federal agencies and a community resilience group affiliated with a National Laboratory[5] who asked the National Research Council to address the broad issue of increasing national disaster resilience (Box 1.3).

BOX 1.3
Increasing National Resilience to Hazards and Disasters
Statement of Task

An ad hoc committee overseen collaboratively by the Committee on Science, Engineering, and Public Policy and the Disasters Roundtable will conduct a study and issue a consensus report that integrates information from the natural, physical, technical, economic, and social sciences to identify ways to increase national resilience to hazards and disasters in the United States. In this context, "national resilience" includes resilience at federal, state, and local community levels. The committee will:

[5] The study sponsors (see also Preface) include U.S. Army Corps of Engineers, Department of Energy, U.S. Department of Agriculture Forest Service, Department of Homeland Security and Federal Emergency Management Agency, National Aeronautics and Space Administration, National Oceanic and Atmospheric Administration, U.S. Geological Survey, and the Oak Ridge National Laboratory/Community and Regional Resilience Institute.

- Define "national resilience" and frame the primary issues related to increasing national resilience to hazards and disasters in the United States;
- Provide goals, baseline conditions, or performance metrics for resilience at the U.S. national level;
- Describe the state of knowledge about resilience to hazards and disasters in the United States;
- Outline additional information or data and gaps and obstacles to action that need to be addressed in order to increase resilience to hazards and disasters in the United States; and
- Present conclusions and recommendations about what approaches are needed to elevate national resilience to hazards and disasters in the United States.

This report responds to this charge by providing actionable recommendations and guidance on how to increase national resilience from the level of the local community, states, regions, and the nation. Because the nation's culture has traditionally been focused on responses to emergencies or to specific disaster events rather than on coherent assessment, planning, and evaluation to increase disaster resilience, the report also recognizes the need for a new national framework for a "culture of disaster resilience" that includes:

(1) Public awareness of and responsibility for managing local disaster risk (Chapter 2);
(2) Establishing the economic and human value of resilience to help encourage long-term commitments to enhancing resilience (Chapter 3);
(3) Tools or metrics for monitoring progress toward resilience and to understand what resilience looks like for different communities (Chapter 4);
(4) Creating local, community capacity, because decisions and the ultimate resilience of our nation derive from the bottom-up community efforts (Chapter 5);
(5) Identifying sound, top-down government policies and practices to build resilience (Chapter 6); and
(6) Identifying and communicating the necessary roles and responsibilities between communities *and* all levels of government in building resilience, including gaps in and challenges to communications and actions among these actors (Chapter 7).

To make the task more manageable, the committee drew from the extensive literature and understanding about natural disasters, but recognizes that many of the ideas and findings are applicable to other hazards and disasters. Chapters 2, 3, and 4 provide a foundation for understanding resilience in terms of management, data, metrics, and approaches that represent important elements of building resilient communities. Chapters 5, 6, and 7 focus on the people—the communities and governing institutions—who make decisions to manage and use data, employ metrics, and implement approaches to help increase resilience. Chapter 8 provides the report's findings and recommendations.

Building and sustaining resilience is everyone's business. Yet, major social and cultural shifts in governance, civility, and trust in institutions such as government, the mass media, and science create barriers that have to be overcome for the nation to move forward. The federal government has already begun a campaign to improve national resilience. PPD-8 states that, "The Secretary of Homeland Security shall coordinate a comprehensive campaign to build and sustain national preparedness, including public outreach and community-based and private-sector programs to enhance national resilience, the provision of Federal financial assistance, preparedness efforts by the Federal Government, and national research and development efforts" (White House and DHS, 2011). True national resilience will integrate these federal efforts with complementary efforts by state and local government, the private sector, and communities at all scales (see Chapter 6 for further discussion of PPD-8).

Inherent in building the culture of resilience is the ability to incorporate scientific information, data, and observing systems to ensure the availability of reliable information, decision support tools, and data sources to decision makers. Enhancing resilience is achieved through vigorous scientific, technical, and engineering research that enables improved forecasting, better risk and disaster management, the development of metrics for assessing progress toward increased resilience, advances in understanding community dynamics, advances in understanding the economics of insurance and disasters, and improved analysis of the legal and social forces at work in communities. Research is essential to building more resilient communities, and research challenges and needs to improve disaster resilience are presented throughout the report.

The report weaves together different kinds of data and experiences from across the nation, including the committee's visits and workshops in the Louisiana and Mississippi Gulf Coasts; Cedar Rapids and Iowa City, Iowa; and Southern California (Appendix B). These examples are used to demonstrate ways in which research in physical and social sciences, engineering, and public health have been tested by the experiences of communities and governing bodies (see also NRC, 2011).

The committee also sought public input through the use of a questionnaire made available through listservs and on the study's Web page.[6] In soliciting information on local opinions across the nation on how resilient their communities are, the committee received both identified and anonymous responses. The quotations that start each chapter are an effort to capture just some of the direct, relevant input the committee received from the wide range of contributors to the study from across the nation. The committee felt that these voices, whether or not they were identified by name, provided thoughtful

---

[6] http://sites.nationalacademies.org/PGA/COSEPUP/nationalresilience/index.htm.

indications of the broad interest in resilience across the country as well as some of its profound challenges.

ON THE NATION'S RESILIENCE AGENDA

This report is viewed as a first step in establishing a vision and national framework for resilience. Fostering a culture of community resilience is viewed as a principal goal for the nation. Building or enhancing resilience at the national level is a long-term process, and it is expected that the tools and framework presented in this report will provide a structure for additional work across communities, including the private sector, and all government levels to advance, measure, and realize resilience in the United States. Enhancing the nation's resilience to disasters will be socially and culturally difficult and politically challenging, and will require certain investments, but the attendant rewards are a safer, healthier, more secure, and prosperous nation. The committee hopes that this report will provide a pathway for achieving this vision for the nation and its communities.

REFERENCES

Bruneau, M., S. E. Chang, R. T. Eguchi, G. C. Lee, T. D. O'Rourke, A. M. Reinhorn, M. Shinozuka, K. Tierney, W. W. Wallace, and D. von Winterfeldt. 2003. A framework to quantitatively assess and enhance the seismic resilience of communities. *Earthquake Spectra* 19(4):733-752.

Colten, C. E., R. W. Kates, and S. B. Laska. 2008. Three years after Katrina: Lessons for community resilience. *Environment* 50(5):36-47.

Cutter, S. L., L. Barnes, M. Berry, C. Burton, E. Evans, E. Tate, and J. Webb. 2008. A place-based model for understanding community resilience to natural disasters. *Global Environmental Change* 18:598-606.

Cutter, S. L., C. G. Burton, C. T. Emrich. 2010. Disaster resilience indicators for benchmarking baseline conditions, *Journal of Homeland Security and Emergency Management* 7(1), Article 51.

DHS Risk Steering Committee. 2008. DHS Risk Lexicon. Department of Homeland Security. Available at www.dhs.gov/xlibrary/assets/dhs_risk_lexicon.pdf.

Flynn, S. 2007. *The Edge of Disaster: Rebuilding a Resilient Nation*. New York: Random House.

Gunderson, L. 2009. Comparing Ecological and Human Community Resilience. Research Report 5. Community and Regional Resilience Initiative, Oak Ridge, TN. Available at http://resilientus.org/library/Final_Gunderson_1-12-09_1231774754.pdf.

Kahan, J. H., A. C. Allen, and J. K. George. 2009. An operational framework for resilience. *Journal of Homeland Security and Emergency Management* 6(1), Article 83. Available at: http://www.bepress.com/jhsem/vol6/iss1/83/.

Mileti, D. 1999. *Disasters by Design: A Reassessment of Natural Hazards in the United States*. Washington, DC: Joseph Henry Press.

Multihazard Mitigation Council. 2005. Natural Hazard Mitigation Saves: An Independent Study to Assess the Future Savings from Mitigation Activities, Vol. 1, Findings, Conclusions, and Recommendations. Washington, DC: National Institute of Building Sciences. Available at ww.nibs.org/MMC/MitigationSavingsReport/Part1_final.pdf.

NCDC (National Climatic Data Center). 2012. Billion Dollar U.S. Weather/Climate Disasters. National Oceanic and Atmospheric Administration. Available at http://www.ncdc.noaa.gov/oa/reports/billionz.html.

NOAA (National Oceanic and Atmospheric Administration). 2011. NWS Central Region Service Assessment: Joplin, Missouri, Tornado—May 22, 2011. U.S. Department of Commerce, National Weather Service, Central Region Headquarters, Kansas City, MO. Available at http://www.nws.noaa.gov/os/assessments/pdfs/Joplin_tornado.pdf.

Norris, F. H., S. P. Stevens, B. Pfefferbaum, K. F. Wyche, and R. L. Pfefferbaum. 2008. Community resilience as a metaphor, theory, set of capacities, and strategy for disaster readiness. *American Journal of Community Psychology* 41:127-150.

NRC (National Research Council). 2011. *National Earthquake Resilience: Research, Implementation, and Outreach.* Washington, DC: The National Academies Press.

NRC. 2012. *Sea-Level Rise for the Coasts of California, Oregon, and Washington: Past, Present, and Future.* Washington, DC: The National Academies Press.

Plodinec, M. J. 2009, Definitions of Resilience: An Analysis. Community and Regional Resilience Initiative, Oak Ridge, TN. Available at http://www.resilientus.org/library/CARRI_Definitions_Dec_2009_1262802355.pdf.

Rose, A. 2009. Economic Resilience to Disasters. CARRI Research Report 8. Community and Regional Resilience Initiative, Oak Ridge, TN. Available at http://www.resilientus.org/library/Research_Report_8_Rose_1258138606.pdf.

Rose, A., and S. B. Blomberg. 2010. Total economic impacts of a terrorist attack: Insights from 9/11. *Peace Economics, Peace Science, and Public Policy* 16(1), Article 2.

Sherrieb, K., F. H. Norris, and S. Galea. 2010. Measuring capacities for community resilience. *Social Indicators Research,* doi10.1007/s11205-010-9576-9.

SDR (Subcommittee on Disaster Reduction). 2005. *Grand Challenges for Disaster Reduction.* Washington, DC: National Science and Technology Council.

UNISDR (United Nations International Strategy for Disaster Reduction Secretariat). 2007. Hyogo Framework for Action 2005-2015: Building the Resilience of Nations and Communities to Disasters. Geneva, Switzerland: UN/ISDR. Available at http://www.unisdr.org/files/1037_hyogoframeworkforactionenglish.pdf.

UNISDR. 2011. Terminology. Available at www.unisdr.org/we/inform/terminology.

White, G. F., and J. E. Haas. 1975. *Assessment of Research on Natural Hazards*. Cambridge, MA: MIT Press.

White House and DHS (Department of Homeland Security). 2011. Presidential Policy Directive-8. Available at http://www.dhs.gov/xlibrary/assets/presidential-policy-directive-8-national-preparedness.pdf.

"We need to recognize that the decision to allow pre-code buildings to stay unretrofitted against local hazards means that the portion of our population that live and work in those buildings face higher than average risks than the populations that are in the newer buildings. [We] need to work to change the dynamic that the areas of town that are most affordable are also the areas facing greater hydrological, geological and ecological risks."

Citizen from King County,
Washington, 2011

2

The Foundation for Building a Resilient Nation: Understanding, Managing, and Reducing Disaster Risks

Understanding, managing, and reducing disaster risks provide a foundation for building resilience to disasters. Risk represents the potential for hazards to cause adverse effects on our lives; health; economic well-being; social, environmental, and cultural assets; infrastructure; and the services expected from institutions and the environment (Figure 2.1). The perceptions of and choices made about risk shape how individuals, groups, and public- and private-sector organizations behave, how they respond during and after a disaster event, and how they plan for future disasters. Most people have some sense of what risk means to them. However, when pressed to identify or assess disaster risk, or determine how to select among available options for managing it, "risk" becomes more difficult to articulate.

This chapter focuses on the importance of understanding risk and risk management as essential steps toward increasing resilience to hazards and disasters. This chapter examines how hazards are identified and how disaster risks are assessed and perceived. Based on this understanding, the chapter summarizes a range of options to mitigate and manage risk. Some of the characteristics of individual and collective decision-making processes—what we know and how we know it—are also described, as are challenges and opportunities that decision makers face in managing risk. Challenges in managing risk due, for example, to inadequate data, to misperceptions of or biases in risk information, to insufficient commitment to use risk management tools, or to lack of communication among stakeholders are also identified. The chapter concludes with several key themes that serve as a foundation for managing risk and increasing disaster resilience for a community, a business, a state, or the nation. Although the chapter directs its discussion of risk and risk management toward general situations using evidence from the published literature, the committee recognizes the importance of the actual practice of risk management. The chapter therefore also draws upon examples from the field and from the standpoint of key decision makers and organizations concerned with addressing disaster risk and increasing resilience.

FIGURE 2.1 Floodwaters rise through downtown Cedar Rapids, June 2008, when the Cedar River finally crested at 31.12 feet, more than 19 feet above the flood stage.
Source: AP photo/Jeff Robertson.

UNDERSTANDING RISK

Disaster risk comprises four elements: hazard, exposure, vulnerability, and consequence (International Bank for Reconstruction and Development/World Bank, 2010) (Box 2.1). *Hazard* refers to the likelihood and characteristics of the occurrence of a natural process or phenomenon that can produce damaging impacts (e.g., severe ground shaking, wind speeds, or flood inundation depths) on a community.[1] *Exposure* refers to the community's assets (people, property, and infrastructure) subject to the hazard's damaging impacts. Exposure is calculated from data about the value, location, and physical dimensions of an asset; construction type, quality, and age of specific structures; spatial distribution of those occupying the structures; and characteristics of the natural environment such as wetlands, ecosystems, flora, and fauna that could either mitigate effects from or be impacted by the hazard.

---

[1] The term "community" throughout the report is used very broadly to incorporate the full range of scales of community organization—from the scale of a neighborhood to that of a city, county, state, multistate region, or the entire nation. Where a specific kind of community is intended, the chapter adds the appropriate descriptor.

Vulnerability is the potential for harm to the community and relates to physical assets (building design and strength), social capital (community structure, trust, and family networks), and political access (ability to get government help and affect policies and decisions). Vulnerability also refers to how sensitive a population may be to a hazard or to disruptions caused by the hazard. The sensitivity can affect the ability of these populations to be resilient to disasters (NRC, 2006b; Cutter et al., 2003, 2008). Vulnerability is projected by the presence and effectiveness of measures taken to avoid or reduce the impact of the hazard through physical or structural methods (e.g., levees, floodwalls, or disaster-resistant construction) and through nonstructural actions (e.g., relocation, temporary evacuation, land-use zoning, building codes, insurance, forecasts, and early warning systems), or construction-related and nonconstruction-related methods.[2]

BOX 2.1
What Is Disaster Risk?

For the purpose of the report, we have adopted a broad definition of risk. The definition presented in this chapter draws common elements from among a range of existing definitions and the communities that provide them. Most definitions take into account elements of *hazard* (what could happen to trigger damage), *exposure* (what is at stake), *vulnerability* (the level of sensitivity to a hazard), and *consequences* (the impact or damage caused by the hazard). We refer to disaster risk as the potential for adverse effects from the occurrence of a particular hazardous event, which is derived from the combination of physical hazards, the exposure, and vulnerabilities (Peduzzi et al., 2009; IPCC, 2012). Similarly, we use the term disaster risk management (or simply risk management) to include the suite of social processes engaged in the design, implementation, and evaluation of strategies to improve understanding, foster disaster risk reduction, and promote improvements in preparedness, response, and recovery efforts (IPCC, 2012).

Consequences are the result of the hazard event impacting the exposure in a region or community, taking into account the degree of the community's vulnerability. Consequences can be immediate (e.g., the loss of human lives, injuries, damaged buildings, businesses), or long term (e.g., environmental

---

[2] The terms structural and nonstructural as they are applied in this report reflect the use of these terms in the flood, hurricane, tsunami, and to a lesser degree, the earthquake arena. Within the emergency management community, the terms are used interchangeably to describe certain mitigation measures. Although the report is consistent in its use of these terms and not outside the norm, nonstructural mitigation has a very specific meaning in engineering circles (it only refers to contents and other building elements not related to structural strength). For the purposes of this report, the committee uses the terms "structural" and "construction-related" and their opposites interchangeably.

damage or physical and mental health impacts), and influence the overall well-being and quality of life for the community (Heinz Center, 2000). Consequences may also extend far beyond the area immediately affected by the hazard—cascading impacts on a supply chain, for example, may have a national or global effect. Lastly, consequences may be mitigated by such measures as insurance, continuity, and recovery plans by businesses and governments, and actions by the state and community such as well-enforced building codes and land-use planning. These measures, put into place either individually or in concert with one another, can greatly reduce the potential losses and facilitate a much speedier recovery from future disaster events, thereby contributing to increasing resilience.

MANAGING RISK

Risk management is a process that examines and weighs policies, plans, and actions for reducing the impact of a hazard or hazards on people, property, and the environment. Ideally, risk is managed in the most effective and equitable way subject to available resources and technical capabilities. Under the best circumstances, risk management includes risk reduction strategies that draw upon scientific, engineering, social, economic, and political expertise. An important aspect of risk management is providing realistic expectations as to what can be accomplished using specific strategies and the relative costs and benefits of undertaking proposed measures (see also Chapter 3). Managing expectations is also important because disaster risks cannot be eliminated completely even with the most appropriate and successful risk management strategies. Importantly also, some tools or actions that can reduce short-term risk may increase long-term risk, requiring careful evaluation of the risk management strategies employed. Although some residual risk will always require attention, risk management can help build capacity to become more resilient to disasters, particularly when everyone in a community is engaged in managing risk (Box 2.2; see also Chapter 5).

The Risk Management Process

Risk management is a continuous process that begins with establishing goals, values, and objectives of the affected and interested parties in the public and private sectors as well as citizen groups and nongovernmental organizations (NGOs) (Keeney, 1992; Sayers et al., 2012) (Figure 2.2). For an affected community, the basis for goal setting begins with questions such as:

- What risks are we facing?
- What risks are we willing to tolerate?
- What risks are not acceptable under any circumstances?

BOX 2.2
Role of Emergency Managers in Risk Management and Disaster Resilience

Although progressive emergency managers anticipate future disasters and take preventive and preparatory measures to build disaster-resistant and disaster-resilient communities, many people are of the opinion that the general field of emergency management does not yet give enough attention to prevention and mitigation activities. Traditionally, emergency managers have confined their activities to developing emergency response plans and coordinating the initial response to disasters. In the future, emergency managers may need to become more strategic in their thinking about disasters in order to help communities respond to the risks they face. The role of the emergency manager necessitates a high degree of technical competence, but is increasingly evolving to include the roles of a manager and a policy advisor who oversee community-wide programs to address risk in all phases of the emergency management cycle. This cycle envelops the characteristics of resilience—to assist communities in preparing and planning for, absorbing, recovering from, and successfully adapting to adverse events. As key actors in risk management and increasing resilience in communities, emergency managers are required to understand how to assess hazards and reduce vulnerability, and to seek the support of public officials and the enforcement of ordinances that reduce vulnerability.

The goals and objectives of the community reflect the values of the key interested parties, current laws, public-sector institutional arrangements at the local, state and federal levels, and existing programs and policies (e.g., the National Flood Insurance Program [NFIP], the California Earthquake Authority, or homeowners insurance offered by the private sector).

Once goals, values, and objectives are established by the nation, state, and/or a community, the next step in the disaster risk management process is *to identify the hazards* (e.g., earthquakes, floods, hurricanes, tornadoes, droughts, ice storms and blizzards, wildfires, landslides, volcanic eruptions, infectious diseases, terrorism, biohazards) and determine whether exposure to them can cause adverse impacts to property, people, and the environment. *Assessing risk*, the next step in the process, is an assessment of the potential impacts associated with these hazards. Risk assessment provides estimates of potential losses to lives and property and some estimate of annual likelihood of occurrence. Sensitivity analysis—part of risk assessment—estimates the efficacy of specific programs and policies in reducing or managing the risk associated with the hazard. *Risk management strategies and decisions* specify the types of information collected by different interested parties in the community and how these data are perceived and used in formulating strategies and programs for managing risk. One of the key factors in *risk strategy implementation* is determining which risks are acceptable or tolerable and which ones are not;

those that are not tolerable thus require management or mitigation (NRC, 2010). The potential consequences of hazards, including losses or disruptions, coupled with the perceptions of risks and consequences play into which risk strategies are used and how they are implemented.

FIGURE 2.2 Continuous and reinforcing process of disaster risk management as a foundation for building resilient communities. Central to the risk management process is the collective evaluation by the community members—including individuals, emergency managers, governing officials, the private sector, and NGOs—of community goals, values, and objectives for the risk management strategy and for community resilience. The entire process, divided for convenience of discussion into six steps, encompasses the ability to identify and assess the local hazards and risks (steps 1 and 2), to make decisions as to which strategies or plans are most effective to address those hazards and risks and implement them (steps 3 and 4), and to review and evaluate the risk management plan and relevant risk policies (steps 5 and 6). The continuity of the process allows a community effectively to "enter" risk management at any point in the "cycle," though identification of basic hazards and assessment of risks is of primary importance.

The last two steps in the disaster risk management process are to *continuously review and evaluate risk strategies* and *to adjust or develop risk management policies*. Although often overlooked, these steps are important, particularly as new opportunities arise, as policies are enacted, or as community goals shift. In designing and evaluating strategies for risk management, new information or data are also important to take into account. Such information may include, for example, knowledge of increased building development in known hazard areas that could increase the exposure to the hazard; the potential impacts of climate change that could affect the intensity or frequency of the hazard; and new and more accurate measurements of key parameters such as precipitation, geological activity along faults, or coastal erosion that influence the way in which a hazard is understood and addressed. Recent disasters in the community or elsewhere can provide lessons and new points of useful information. By recognizing and reviewing risk strategies and available (and sometimes new) data on hazards and their impacts, adjustments can be made to overcome deficiencies and improve the existing set of policies, institutional arrangements, and strategies to develop new ones, allowing the risk management cycle to begin again. Emergency managers use risk management principles described in this cycle to establish priorities for the communities within their jurisdiction (Box 2.3).

BOX 2.3
Emergency Managers as Risk Management Practitioners

The following is extracted from the document "Principles of Emergency Management" (IAEM, 2007) and identifies some of the principles of emergency management that relate to the role of emergency managers as practitioners of risk management:

Emergency managers generally employ risk management principles such as hazard identification and risk analysis to identify priorities, allocate resources and use resources effectively. . . . Setting policy and programmatic priorities is therefore based upon measured levels of risk to lives, property, and the environment. The National Fire Protection Association (NFPA) 1600 states that emergency management programs should identify and monitor hazards, the likelihood of their occurrence, and the vulnerability to those hazards of people, property, the environment, and the emergency program itself. The Emergency Management Accreditation Program (EMAP) Standard echoes this requirement for public sector emergency management programs. . . . Emergency managers are seldom in a position to direct the activities of the many agencies and organizations involved in emergency management. In most cases, the people in charge of these organizations are senior to the emergency manager, have direct line authority from the senior official, or are autonomous. Each stakeholder brings to the planning process their own authorities, legal mandates, culture and operating missions. The principle of coordination requires that the emergency manager, or other actors responsible for risk management and increasing resilience, gain agreement among these disparate agencies as to a common purpose, and then ensure that their independent activities help to achieve this common purpose.

Note: Information on NFPA 1600 is available at http://www.nfpa.org/newsReleaseDetails.asp?categoryid=488&itemId=46745&cookie%5Ftest=1; EMAP information is available at http://www.emaponline.org/.

Foundation for Risk Management

Two elements provide the foundation for managing risks: identifying the hazards that affect the community and assessing the risks that such hazards pose (see Figure 2.2). Both are based on scientific information. Because these two steps provide the cornerstone for risk management, we provide more detail on the current methods for hazard identification and disaster risk assessment.

Hazard Identification

As noted earlier, *hazard identification* determines the types and characteristics of potential disasters facing a community or region (Box 2.4). For example, earthquake hazard is a combination of the likelihood of earthquake occurrence (location, magnitude, and recurrence rate of all future damaging earthquakes impacting a region) and ground motion predictions that are used to calculate the spatial distribution of shaking intensity for these future events. In a similar way, a hurricane hazard can be described by the spatial distribution of its projected path and wind speed and central pressure along that path. Assessing the likelihood of earthquake- and weather-related events typically is based on analysis of - both the historical and geological record of events, knowledge of the physical processes leading to the occurrence of a disaster, and real-time data collection and monitoring of natural (geological, atmospheric, oceanic) phenomena. Although historical records are important, limits exist on the extent to which generalizations can be made about how physical phenomena will evolve in the future. For example, expected changes in climate bring into question how to interpret historical data in characterizing the intensity and magnitude of future hurricanes and floods (Milly et al., 2008), and may increase the costs and losses associated with severe storms and extreme events in the years to come (Karl et al., 2009; NRC, 2011a; IPCC 2012).

BOX 2.4
Cedar Rapids, Iowa: Hazard Identification

In May and early June 2008, tornadoes and floods struck Iowa. The largest single tornado in the state in a 30-year period, an EF-5,[a] struck the town of Parkersburg, Iowa, 85 miles northwest of Cedar Rapids on May 25 and caused millions of dollars in damage, eight deaths, and the mobilization of significant state and local emergency response resources.

In early June, as the effects of the tornadoes were still being evaluated and absorbed, the residents and decision makers of Cedar Rapids were monitoring information about the potential for major flooding of the Cedar River which passes through the city center. The water levels in the Cedar and nearby Iowa Rivers and their tributaries had risen throughout the spring because the agricultural land that covers 74 percent of the state, still saturated from the heavy winter snowmelt and without crop cover, together with an extensive network of subsurface clay drainage tile systems, contributed extensive runoff into the rivers. The high river levels were exacerbated by heavier-than-average precipitation during the spring (Bradley, 2010; Krajewski and Mantilla, 2010). Having endured record floods in 1993 when the Cedar River crested in Cedar Rapids at 22.5 feet (the river's flood stage is 12 feet), most citizens, officials, emergency personnel, businesses, and museums held some expectation that they would not risk another "100-year flood" in 2008. When the Cedar River eventually crested (see Figure 2.1) at more than 31 feet, it was well above what would characterize a "500-year" flood event.[b]

Hazard identification is more than just historical experience with hazard events; it includes the identification of potential sources of disaster to the community and the likelihood and expected impacts of future events. Cedar Rapids has multiple sources of natural hazards: floods, severe weather (thunderstorms and hail; severe winter weather), tornadoes and severe wind storms, and heat waves. Cedar Rapids (Linn County) is also located 9 miles downstream from the Duane Arnold Energy Center, a commercial nuclear power facility, and is within the emergency planning zone for that facility, adding a direct human-made hazard to the area.

The city and county have a risk mitigation strategy in place for the nuclear power facility: the city's emergency planners, hospital personnel, and citizens drill four times a year along established evacuation routes. These drills, including the relocation of essential medical facilities and personnel proved essential during the response to the flooding of the Cedar River into the city in the second week of June 2008. According to the health personnel and emergency responders with whom the committee spoke in their visit to Cedar Rapids, the preparation and planning involved in preparing for that single, human-induced hazard played a large role in the fact that no lives were lost to a different hazard that evolved into a disaster during the flooding in 2008.

[a] "EF" equates to the Enhanced Fujita scale, which is a tornado rating based on estimated wind speeds and damage. The scale ranges from EF-0 to EF-5. At EF-5, wind speeds are estimated to exceed 200 mph for 3-second gusts (http://www.crh.noaa.gov/arx/efscale.php).

[b] The 100-year floodplain is the boundary of the flood that has a 1 percent chance of being equaled or exceeded in any given year; the 500-year floodplain has a 0.2 percent chance of being equaled or exceeded in any given year.

Sources: Panelists in the committee's field trip and workshop in Cedar Rapids and Iowa City (see Appendix B for list of panelists); www.linncounty-ema.org; http://www.crh.noaa.gov/Image/dmx/Iowa%20Tornado%20Statistics%201980-2008%20Graph.pdf.

Data and characterization of weather-related events and other natural hazards such as earthquakes, floods, or wildfires are made by federal agencies such as the U.S. Geological Survey (USGS), National Oceanic and Atmospheric Administration (NOAA), Federal Emergency Management Agency (FEMA), U.S. Army Corps of Engineers (USACE), National Aeronautics and Space Administration (NASA), and U.S. Forest Service, each of which has responsibility for collecting data and monitoring these phenomena (see Chapter 6 for more detailed description of these federal roles and Appendix C for some of the kinds of data that these agencies collect and monitoring that they conduct). Much of this information is provided to communities in data tables or in the form of maps.

One example of a quantitative hazard assessment for a specific hazard is well illustrated by the USGS National Seismic Hazard Mapping Project, recognized both nationally and internationally as the authoritative analysis of earthquake hazard in the United States. The USGS process includes solicitation of input parameters from regional experts, a logic-tree approach to capture the range of scientific uncertainty in input parameters, transparency regarding all input data and methodology, and online accessibility to a wide array of digital hazard maps and derivative products.[3] One of the major strengths of the USGS Seismic Hazard Maps is that they are probabilistic; that is, they provide estimates of ground-shaking levels at different return periods for the full array of potential future earthquakes and take into account each earthquake's rate of occurrence.

Many communities address their potential hazards in a qualitative way, such as by defining high-, moderate-, and low-hazard zones, or through scenarios of likely or worst-case events, but only a probabilistic hazard assessment quantitatively captures potential events and their impacts together with their likelihood of occurrence. Probabilistic hazard assessment draws from historical data but also from longer-term records of past events from the geological record. The USGS's probabilistic hazard is used to develop outputs of earthquake ground motion for designing buildings and structures that accord, for example, with the 2012 International Building Code.[4] For example, most building codes in the United States are based on the USGS's estimate of the ground motion level with a 10% probability of exceedance in 50 years. This corresponds to ground motions with a 475-year return period, or the highest shaking level expected from any nearby earthquake source that is likely to occur over the next 475 years.[5] Probabilistic hazard is also the input used in risk assessment to compute probable losses at different return periods and is thus used to determine insurance premiums for relatively low likelihood but high-impact events.

The largest federal hazard mapping program is NFIP's flood insurance rate maps, produced for the community level. These maps identify areas subject to flooding from events of varying intensity based on elevation, channel morphology and streamflow, and watershed conditions. Elevation data are based on topographic features using digital elevation models. The flood risk information is based on hydrological and hydraulic analyses, historical data, and

---

[3] See http://earthquake.usgs.gov/hazards/.

[4] https://geohazards.usgs.gov/secure/designmaps/us/.

[5] See USGS FAQs: http://earthquake.usgs.gov/learn/faq/?faqID=223.

watershed characteristics as they affect runoff. Although the flood mapping process and inputs are well known, actually making accurate flood maps and maintaining the information are complex (NRC, 2009). Limitations in our understanding of floodplain boundaries, needed improvements in predictive and probabilistic flood models (riverine and storm surge), and enhanced topographic accuracy (NRC, 2007a, 2009) render the timely production of flood maps a costly but essential proposition for communities and the federal government (Box 2.5).

BOX 2.5
A College Campus Benefits from Flood Maps After Hurricane Irene (2011)

The Russell Sage College Campus in Troy, New York, sits within two blocks of the Hudson River, north of Albany. On August 28, Hurricane Irene had passed through the area. Although Monday, August 29 was clear and sunny, the Hudson River was rising. The disaster management team at the college used FEMA flood maps to estimate the risk of campus flooding, which would necessitate the evacuation of all personnel and students who had just arrived to begin the fall semester. Although the start of the academic year had to be delayed, the river stopped rising just below the level at which the campus would have flooded. Only the basements of two low-lying buildings were affected. The flood maps were essential in preventing an unnecessary evacuation.

In some states, the federal and state agencies work together to develop authoritative zoning maps to identify areas subject to multiple levels of hazards for a variety of perils such as landslides, liquefaction, and surface fault rupture. Also, new technologies are making possible increasingly higher resolution and more sophisticated and detailed hazard identification maps such as the characterization and monitoring of wildfire activity (Figure 2.3).

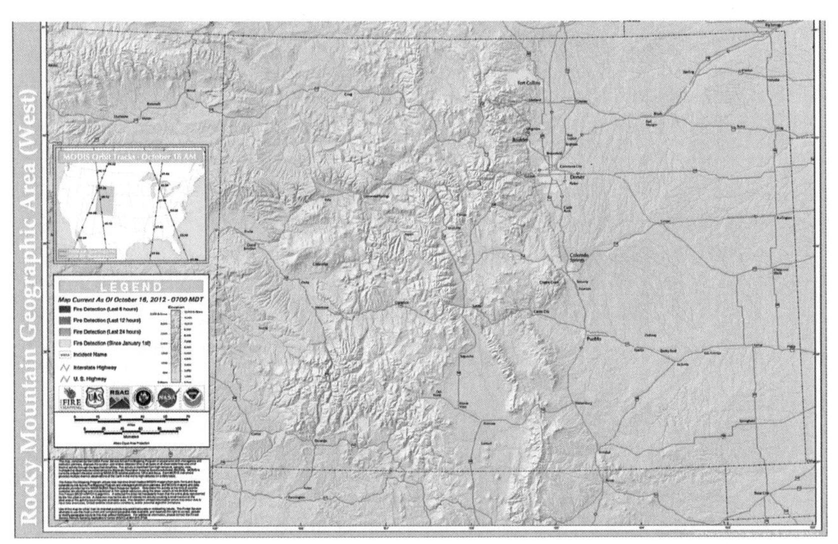

FIGURE 2.3 Active wildfire map of part of the Rocky Mountain area showing wildfires (yellow) from the U.S. Forest Service Remote Sensing Applications Center. The Active Fire Mapping Program is a satellite-based fire detection and monitoring program that provides near real-time detection and characterization of wildland fire conditions for the continental United States, Alaska, Hawaii, and Canada. Detectable fire activity in the United States and Canada is mapped and characterized by the program. High temporal image data collected by NASA's Moderate Resolution Imaging Spectroradiometer (MODIS) are the primary remote sensing data source of this program at present. MODIS provides multiple daily observations of the United States and Canada, which is ideal for continuous operational monitoring and characterization of wildland fire activity. Such data and maps are essential for those fighting the fires, as well as for city and town officials, and individual homeowners. Source: U.S. Forest Service, Remote Sensing Application Center.

Risk Assessment

The risk assessment process combines the physical characteristics of potential hazards obtained through hazard identification with data on exposure, vulnerability, and mitigation measures. Risk assessment involves estimating the likelihood of specific events occurring, their potential consequences, and the uncertainties surrounding these estimates.

At the simplest level, a community can overlay maps of high, medium, and low hazard (as described above) on maps of exposure (properties at risk) to estimate disaster risk. A more rigorous approach would include an additional layer of structural vulnerability (susceptibility to damage from impacts from that hazard) to determine the riskiest regions in a community (those with high vulnerability, large exposure, and high hazard) and the effects of mitigation measures. Many communities now use geographic information systems (GIS) to map the location, type, and value of community assets. GIS provides the ability to store and manage vast amounts of spatially referenced information and thus has become an ideal environment for conducting cost-effective hazard and risk assessments.

Risk assessment was greatly improved by the confluence of two developments in the last several decades: development of scientifically based probabilistic hazard assessment (quantifying the rate of occurrence and magnitude of hazard events and their impacts; Cornell, 1968) and advances in information technology and GIS (Cutter, 2001; NRC, 2007b; Emrich and Cutter, 2011). Taking advantage of these developments, a new risk management industry developed in the late 1980s and early 1990s and created computer-based models for quantifying probabilistic catastrophe risk and loss potential at different return periods. These so called probabilistic "cat models" (catastrophe models) now form basis for determining premiums for natural hazard insurance (Grossi and Kunreuther, 2005).

FEMA has produced a freely available catastrophe modeling tool, HAZUS, to provide communities with the capability to run scenarios or actual events (earthquake, flood, and hurricane wind) impacting the community in order to estimate losses (e.g., property damage, casualties, infrastructure disruption, and displaced households) for planning or post-disaster recovery operations (see Chapter 3, Figure 3.4 for HAZUS example).[6] Catastrophe models such as HAZUS present an opportunity for community leaders, regulators, and emergency management agencies to design risk management strategies by comparing potential losses with and without mitigation measures in place for specific scenarios, so called deterministic (as opposed to probabilistic) risk. Such applications require upgrading of the default HAZUS building and infrastructure inventories to get meaningful local loss results. Additional improvements in the direct and indirect economic loss modules would also be relevant to translate these losses into business interruption losses for direct customers and indirect losses up and down the supply chain. A HAZUS study for an earthquake scenario that involved an earthquake of magnitude 7.7 striking in the middle of the country near New Madrid, Missouri, was recently released. The study invested considerable effort in populating the public infrastructure database in HAZUS and determining appropriate infrastructure fragility relationships in order to more accurately determine potential impacts to the infrastructure network (Elnashai et al., 2009). Similarly, the state of North Carolina is in the process of developing detailed exposure data on the inventory (location and construction type) for all structures in all communities as a means for improving the accuracy of the input data for the HAZUS loss model.

---

[6] http://www.fema.gov/hazus/.

DECISION MAKING UNDER RISK AND UNCERTAINTY

Decisions on risk-reduction and mitigation strategies are a function of the roles and responsibilities of decision makers, the influences on these decision makers, and the policy options available to them. Given different backgrounds and inherent conditions, communities faced with the same challenge may develop entirely different portfolios of risk reduction measures. As mentioned previously, actions that can reduce short-term risk can potentially increase long-term risk. For example, elevating homes in a coastal area above currently predicted storm surge levels may encourage continued development in an area that is subject to a variety of other hazards such as wind storms, coastal erosion, flooding, and hurricanes for which home elevation alone may not adequately reduce the long-term risk. Another example, detailed later in this chapter is building of levees or other structures that are designed to prevent floodwaters, storm surges, or other hazards from reaching areas that are at risk. In the short term, the presence of the levee may reduce risk to the local hazard; however, if the upper limit to the capability of these structures is ever exceeded, the consequences to those with homes or businesses behind the levee can be catastrophic (Tobin, 1995; Burby, 2006; Cutter et al., 2012; NRC, 2012). Decision making for risk management that helps to increase disaster resilience includes analysis of costs and potential benefits; the significant challenges lie in recognizing that benefits, whether economic or otherwise, are not necessarily equally distributed among those incurring the costs. This topic is explored in more detail in Chapter 3.

The key actors in managing disaster risk include the public sector at local, state, and federal levels who conduct and design hazard management programs; residents and businesses in hazard-prone areas; those who provide ways to mitigate losses prior to a disaster (e.g., developers, insurers, banks, and NGOs); those who provide services to those affected by the disaster during the recovery period (e.g., emergency managers, fire, police, hospitals, and NGOs); and the research community that provides analysis of risk, hazards, and disasters. Some of the responsibilities, challenges, and opportunities facing these key interested parties are captured in Table 2.1.

TABLE 2.1 Responsibilities, Challenges, and Opportunities of Key Interacting Parties in Risk Management

Interested Party	Responsibility	Challenges	Opportunities
Federal government	Provides, and in some cases operates, protection structures for communities; supports NFIP; provides disaster assistance	No comprehensive or coordinated approach to disaster risk management	Stemming the growth in outlay of post-disaster recovery funds
State and local governments	Ensure public health and safety in use of land, zoning, land-use planning, enforcement of building codes, development of risk management strategies	Reluctance to limit development; difficulty in controlling land use on privately owned land	Reaping benefits of multiple ecosystem services by investing and strengthening natural defenses
Homeowners and businesses in hazard-prone areas	Take action to reduce vulnerability and increase resilience of property	May be unaware of or underestimate the hazards that they face	Creating demand for disaster-resistant or retrofitted structures that have increased value
Emergency managers	Oversee emergency preparedness, response, recovery, and mitigation activities	More focused on immediate disaster response than risk management	Reorientation of training and roles to balance focus toward prevention and overall disaster resilience
Construction and real estate	Incorporate resilience into designs; inform clients of risk	Actions may increase cost and reduce likelihood of sales	New opportunities in niche market

TABLE 2.1 continued.

Interested Party	Responsibility	Challenges	Opportunities
Banks and financial institutions	Require hazard insurance	No incentives to require insurance	Reduce overall risk in their portfolios
Private insurers and reinsurers	Offer hazard insurance at actuarial rates; identify risks	Limits may be placed on rate structures	Greatly expanded and risk-reduced market by offering incentives such as premium reductions for retrofit measures
Capital markets	Catastrophe bonds[a] and other alternative risk transfer instruments	Availability limited due to globalized financial markets	Large resource base and new investment opportunities that could be directed in an anticipatory way
Insurance rating agencies	Identify stability of insurers	May negatively impact insurer position	Transparency to enable informed decisions on the part of consumers
Researchers	Collect, analyze, and communicate data, forecasts, and models about risk, hazards, and disasters	Insufficient or dispersed datasets; understanding how to share scientific information with broad audiences	Increased forecasting capability and improved data-based models of physical processes leading to disasters

[a]Catastrophe bonds ("cat bonds") are risk-linked financial tools that can be used by insurance companies to cover the potential risk of a major catastrophe and the premiums that would have to be paid by the insurance company if the disaster or catastrophe was to occur. Insurance companies are required by state law to have capital on hand to cover routine losses, but for higher losses, they may buy reinsurance or issue cat bonds. The insurance companies issue cat bonds through an investment bank to investors. The insurance company will use the funds from the bond issuance to pay insurance claims if a catastrophe occurs; if the catastrophe does not occur within a specified time interval (usually some number of years), the insurance company pays the amount of the bond with interest to each investor. See Kunreuther and Michel-Kerjan (2011, Chapter 8).

bank to investors. The insurance company will use the funds from the bond issuance to pay insurance claims if a catastrophe occurs; if the catastrophe does not occur within a specified time interval (usually some number of years), the insurance company pays the amount of the bond with interest to each investor. See Kunreuther and Michel-Kerjan (2011, Chapter 8).

Empirical and experimental research by economists, geographers, psychologists, and other social scientists reveals systematic biases that decision makers pursue with respect to their perception of risk, their experiences, and the actions they choose to take in advance of a disaster and after an event occurs (Slovic et al., 1978; Slovic, 1987, 2000; Kunreuther et al., 2012). Those behavioral features most influential in the development of risk management policies include (see also Box 2.6):

- *Risk perception.* Psychological and emotional factors that define risk perception have an enormous impact on behavior (Magat et al., 1987; Huber et al., 1997; Slovic, 2000).
- *Status quo bias.* There is a tendency to maintain current behavior rather than seek new options. If given an opportunity to postpone an investment for a month or a year, there will be a tendency to delay the outlay of funds (Samuelson and Zeckhauser, 1988).
- *Myopic behavior.* There is a tendency for individuals to be myopic when making decisions with respect to preparing for disasters. By focusing on short-term returns, they fail to invest in risk-reducing measures that could be justified financially when comparing costs and expected returns over the expected life of the property (Kunreuther et al., 2012).
- *Simplified decision rules.* In making choices with respect to protection against low-probability risks such as natural disasters, individuals often use decision processes that involve simplified heuristics and rules of thumb rather than undertaking more systematic evaluations of alternatives such as rigorous benefit-cost analyses (Camerer and Kunreuther, 1989; Kahneman and Tversky, 2000).
- *Reframing likelihood data.* Communicating risk information is of fundamental importance but such information is not always successfully transmitted to decision makers who most need it. By reframing information on the likelihood of an extreme event occurring, it may capture the attention of decision makers rather than being below the threshold level of concern. For example, if a flood with a return period of 100 years was presented as having a greater than one in five chance of occurring in the next 25 years, then key stakeholders may have an interest in taking steps to reduce the potential losses (Kunreuther et al., 2001; Galloway et al., 2006).

BOX 2.6
Lessons from the Field: Behavioral Basis for Decision Making

Risk perception and purchasing flood insurance: Personal experience with flooding in Cedar Rapids led to those residents closest to the river to take measures to protect their property by purchasing flood insurance, by moving their possessions off the ground floor, and sandbagging. However, on the west side of the city, which has a higher percentage of elderly, lower-income, and disabled residents, only a limited number of homes (information shared with the committee indicated about 10 percent of residents) had flood insurance (Figure).

FIGURE. One of the homes on the neighborhood on the west side of the Cedar River in Cedar Rapids that was inundated by floodwaters. Although many homes have been rebuilt to more flood-resistant standards in this neighborhood through concerted community and city effort, many homes remain damaged and uninhabited. Cedar Rapids continues to perform more than 1,200 acquisitions and has demolished about 900 structures. More than 200 structures remain damaged and uninhabited.
Photo source: John. H. Brown Jr./The National Academies.

Status quo bias: This bias is illustrated by the relative lack of preparedness demonstrated by the city of New Orleans and FEMA in advance of Hurricane Katrina in 2005. Two months prior to the storm, the city conducted a full-scale simulation that demonstrated what would happen if a hurricane of Katrina's strength struck (Brinkley, 2006). As the active hurricane season approached, little was done to remedy known flaws in their preparedness plans.

strength struck (Brinkley, 2006). As the active hurricane season approached, little was done to remedy known flaws in their preparedness plans.

Simplified decision rules bias: In Waveland, Mississippi, town officials acknowledged they were not prepared for a storm event of Katrina's magnitude. No event of that magnitude had ever occurred in Waveland that might have allowed residents or decision makers to anticipate the ultimate effects of the storm. A railroad embankment had for years protected homes north of it in previous hurricanes, so residents behind the embankment felt less need to evacuate. As a result of the decision not to evacuate, fatality rates were higher in Waveland than elsewhere on the coast (NRC, 2011b).

Reframing likelihood data: As the Cedar River rose in June 2008, federal government agencies provided updated information and data on the change in the river level and the weather—the National Weather Service (NWS), the U.S. Geological Survey (USGS; which maintains stream gages[a]), and the U.S. Army Corps of Engineers (USACE, which maintains the levee system and regional reservoirs) exchanged updates with one another, with city officials and emergency personnel in Cedar Rapids, and with state emergency and health personnel. In response to these updates and other information on conditions in the city, the local and state personnel communicated and continuously reframed their plans and responses as the river level climbed and the likelihood of a large flood became evident.

[a]Stream gages measure and record river stages (height) and send the data in real time to a central office and over the Internet. These stream gages are distributed across the nation, and are maintained by the USGS, and are usually funded cooperatively with state or local governments. These data are provided to other government agencies such as the NWS and USACE.

RISK MANAGEMENT STRATEGIES AND MEASURES

Risk management strategies often represent the integration of structural and nonstructural measures designed to reduce vulnerability and mitigate consequences. In some cases, the risk management strategy consists primarily of structural (construction-related) measures such as levees, floodwalls, or disaster-resistant construction and retrofitting. Other strategies may focus on nonstructural (nonconstruction-related) means such as land-use management and planning, utilizing natural defenses such as swamps and wetlands (green infrastructure) to reduce the impact of flooding on communities, building codes, insurance, early warning systems, and evacuation. In most cases communities include portfolios of both structural and nonstructural measures; the combination of these measures improves the likely success of reducing hazard impacts and also improving resilience.

In evaluating alternative measures for managing risks associated with natural disasters and making the community more resilient, decision makers need to comprehensively evaluate the advantages and disadvantages of each measure and the possible impacts on different individuals and groups residing in the community over time (Box 2.7). For example, a strategy that appears to mitigate some of the risks effectively (e.g., not allowing homes or businesses to be rebuilt on a floodplain) may have an adverse impact on one or more social groups and would therefore be deemed unacceptable by some members. Another proposed measure could create long-term fiscal commitments for the community that cannot be met without additional taxes that would be deemed by most residents to be unacceptable. Finding consensus among these needs and requirements is essential to implementation of any risk management strategy. The subsequent paragraphs provide a brief description of the principal structural and nonstructural measures currently being employed. Table 2.2 highlights typical actors, time frames, and potential benefits and adverse impacts of the structural and nonstructural measures and tools discussed below.

BOX 2.7
Determining Costs and Benefits of Different Management Strategies

Once the hazard is identified and assessed, decision makers can determine what strategies they will employ to reduce the risk the community faces. In developing these strategies officials need to estimate the benefits and costs of different measures as well as determining who should pay for them. If a probabilistic risk assessment has been carried out, there is a sound actuarial basis for analyzing the cost-benefit analysis of potential risk reduction/mitigation measures. Insurance can play a key role in encouraging the adoption of these measures. Suppose a family could invest $1,500 to strengthen the roof of its house so as to reduce the damage by $30,000 from a future hurricane with an annual probability of 1/100. An insurer charging a risk-based premium would be willing to reduce the annual charge by $300 (i.e., 1/100 x $30,000) to reflect the lower expected losses that would occur if a hurricane hit the area in which the policyholder was residing. If the house was expected to last for 10 or more years, the net present value of the expected benefit of investing in this measure would exceed the upfront cost at an annual discount rate as high as 15 percent so that the measure would be deemed to be attractive and viewed as cost-effective. If the homeowner could obtain a $1,500 home improvement loan tied to the mortgage at an annual interest rate of 10 percent, this would result in payments of $145 per year. Assuming that the insurance premium was reduced by $300, the savings to the homeowner each year would be $155 ($300-$145).

Structural (Construction-Related) Mitigation

The effect that hazards have on an exposed area can be mitigated by structural systems put in place to reduce the effects of the event. Some of these are locally developed, owned, and operated, while others require negotiated partnerships and joint decisions between local, state, and federal interests, and resources (Figure 2.4). A brief description of some of the most frequently used measures follows.

FIGURE 2.4 Structural Flood Mitigation Measures, clockwise from top left: Mississippi River levee; Grand Coulee Dam on the Columbia River in Washington; a floodwall that protects the city Winona, Minnesota, along the Mississippi River; and the Bonnet Carré Spillway, a floodway that diverts water from the Mississippi River into Lake Ponchartrain to reduce the flow passing New Orleans.
Sources: USACE (Mississippi River Levee) http://www.mvd.usace.army.mil/mrc/mrt/Docs/Levees%20info%20paper.pdf; USACE (Winona floodwall) https://eportal.usace.army.mil/sites/DVL/DVL%20Images/cemvp191.tif; USBR (Grand Coulee Dam) http://www.usbr.gov/pn/grandcoulee/; http://users.owt.com/chubbard/gcdam/html/photos/exteriors.html; and NASA (Bonnet Carré Spillway) http://earthobservatory.nasa.gov/IOTD/view.php?id=87381.

Levees, floodwalls, and similar structures

These structures are designed to prevent floodwaters, storm surges, or lava flows from reaching areas that are at risk. When the upper limit to the capability of these structures is exceeded, the consequences can be catastrophic. For example, when levees overtop or fail, those people behind the barriers are subjected to conditions more severe than they would have had if the flood or lava flow had been more gradual in its approach. The ways in which dam and levee safety can be integrated with community resilience have been discussed in a recent report (NRC, 2012).

Dams and flood control

Dams retain flood waters before they reach an area at risk. Some of the pool behind a dam is set aside to store floodwaters during high-precipitation periods and then release the stored water gradually to reduce the likelihood of damage to the community at risk. This storage can be inadequate if there are exceptional rainfall and snowmelt events and as was illustrated by the 2011 rain events on the Missouri River system. At some point, dams may no longer be able to contain the waters, and uncontrolled flows move downstream and either add to the existing flood or initiate flooding. On rare occasions, dams can fail and may inundate those below. Dam failures can cause significant property losses and environmental damage. State dam safety programs, for example, reported 132 dam failures in the period from 2005 to 2008,[7] although only one of these failures resulted in loss of life, in part because most of the dams were of limited size.

Floodways

Where the capacity of a river to pass a large volume of flood waters through a critical location is limited, floodways, spillways, or channels are constructed to carry these flows around the community or region. In the 2011 flooding along the Mississippi River, USACE relieved downstream flooding near Cairo, Illinois, by breaching upstream levees and flooding agricultural fields on leased land that had been held in reserve for exactly this purpose.

Disaster-resistant construction and retrofitting existing building stock

A significant opportunity to reduce loss in future events and thus increase resilience is to strengthen and/or retrofit the nation's existing building stock. In the case of hurricanes, the new construction and retrofitting is relatively inexpensive and can include installation of exterior hurricane shutters

---

[7] http://damsafety.org/newshttp://damsafety.org/news/?p=412f29c8-3fd8-4529-b5c9-8d47364c1f3e.

or replacing windows with impact resistant glass, garage door bracing, strengthening soffits, and securing loose roof shingles. In portions of the nation with high seismic hazard, strengthening older and structurally weak construction can require modest (in the case of wood frame one- to two-story structures) to substantial (in the case of 1960s and earlier nonductile concrete frame construction) investments. For structures that need to function immediately after an earthquake (such as hospitals, city halls, emergency operations centers) base isolation consisting of shock-absorbing devices that help isolate the building from strong ground shaking greatly reduces the possibility of damage; however, such strategies can be very expensive.

Hazard-conscious ("Smart") building

Individual structures can be elevated, flood proofed or constructed to resist most hazard forces in order to reduce losses from future events such as floods, hurricanes, windstorms, and earthquakes (FEMA, 1998). "Smart" buildings can adjust to certain changes in conditions to counteract damaging structural reactions in response to an external hazard.

Securing building components and contents from damage from shaking, strong winds, or flooding

The failures of nonstructural components during earthquakes and other disasters may result in injuries or fatalities, cause costly property damage to buildings and their contents, and disrupt the operation of or force the closure of residences, businesses, and government offices. Bachman (2004) suggests that the nonstructural component and building content losses in recent events in developed countries represent 50 percent of total earthquake losses, but it is difficult to find sufficient data to substantiate this view.

Well-enforced building codes

Building codes can be adopted at the state or local level, but require local enforcement. Hurricane Andrew in 1992 revealed that one-third of the damage could have been avoided had Florida enforced its building codes (Kunreuther, 1996). Public officials may exacerbate the problem by not enforcing building codes and/or imposing zoning restrictions. See detailed descriptions in Chapter 5.

Nonstructural (Nonconstruction-Related) Mitigation and Risk Transfer

Nonstructural measures span a range of activities including securing building components and contents from damage due to strong shaking, winds, or floods; timely and accurate forecasts and warning systems; locally based changes in zoning and land use; and improved communication of risks. In many instances, nonstructural measures for disaster risk reduction necessitate local control of decisions and implementation, although state and federal partnerships help support the programs. For example, U.S. residents can purchase flood insurance through the federally run NFIP when the communities in which they live agree to participate in the program. The program requires the community to adopt the flood insurance rate maps and to adopt and enforce floodplain management ordinances and control land-use development in the 100-year floodplain. Over 21,000 communities currently participate in the NFIP.

Natural defenses

Many types of natural defenses against disasters exist. For example, wetlands and swamps can store overflow waters from riverine flooding and help reduce downstream impacts; wetlands also provide a natural barrier to storm surge inundation (Galloway et al., 2009). Coastal sand dunes protect structures built behind them and help slow coastal erosion and also the immediate impact of rising storm surge. Communities and regions may include enhancement of these natural features to improve their risk reduction capabilities (Opperman et al., 2009).

Risk mapping

Combining the natural hazard risk assessment with quantitative consideration of mitigation measures yields expected outcomes that can be graphically portrayed in a manner that facilitates public understanding of the risk and its implications for them. Critical to risk mapping as a tool to manage risk is that the information is properly communicated to those who need to use the data. Risk communication is discussed in detail in Chapter 5.

Zoning ordinances

Zoning policies are locally controlled and enforced and can prohibit building or rebuilding in hazard-prone locations. See Chapter 5 for details on zoning and building codes.

Hazard and vulnerability disclosure

Everything that a buyer needs to know about a new car is on the Monroney sticker (United States Code, Title 15, Chapter 28, §§ 1231-1233). The Monroney sticker is required in the United States by federal law for new cars and includes, in addition to make, model, and serial number, items such as the final assembly point of the car, the manufacturer's suggested retail price, optional equipment, safety ratings, and acknowledgment if the car has not been tested for safety. Except for disclosure mechanisms that have been legislated in a few states to inform potential home buyers that the property they are buying is located in the pathway of a potential hazard, the real estate industry's multiple listing service (MLS) is not required to provide information on the structural integrity of the house or its location with respect to nearby hazards. The MLS is not required to give any information about the roof construction and its tie-downs, for example, and it does not indicate if the home was built to code (either at the time of original construction or whether it meets present codes), or if additions or retrofits have been made by a licensed contractor or by someone who is not licensed.

In California, for example, zones of potential landslide, liquefaction, or fault rupture hazard have been mapped by the California Geological Survey as "special study zones" according to provisions in the California Alquist-Priolo Earthquake Fault Zoning Act of 1972.[8] If a property is in one of these special study zones, the buyers must sign a form indicating that they have been made aware of this potential hazard and recognize that additional inspections and work may be required if they choose to modify the property in the future.

The U.S. Resiliency Council,[9] a nonprofit organization, is working on creating building "report cards." They are developing technically defensible metrics to evaluate and communicate the resilience of individual buildings. Their initial focus is on seismic risk, and they later plan to extend their efforts to creating metrics for resilience to catastrophic wind and flood risk. Transparency and required disclosure of these individual building resilience ratings will benefit building users, owners, and lenders by increasing the value of well-designed or properly retrofitted properties. Policy makers will be able to use

BOX 2.8
Property Transfer Tax Program, Berkeley, California

The Property Transfer Tax Program in Berkeley has provided funds for seismically retrofitting a huge number of properties in the city. In 1992, voters approved an additional 0.5 percent transfer tax on top of the existing 1 percent tax on all real estate transactions, with the tax paid equally by buyer and seller. This 0.5 percent portion of the transfer tax is available for voluntary seismic upgrades to residential property. Residential property owners have up to 1 year to complete the seismic retrofit (or lose the funds). Since many homes sell for $750k to $1M or more in Berkeley, this amounted to $3,750-5,000 in "free funds" and can cover homeowner upgrades such as brick chimney bracing or anchoring water heaters. This incentive program has an 80-90 percent participation rate. Along with other measures, this program has led to more than 60 percent of the residences in Berkeley becoming more seismically resistant.

SOURCES: http://www.ci.berkeley.ca.us/ContentDisplay.aspx?id=6282; http://www.eeri.org/mitigation/files/berkeley.transfer.tax.rebate.pdf.

---

8 http://ceres.ca.gov/planning/pzd/2000/pzd2000_web/pzd2000_misc19.html.

9 http://www.usrc.org/.

ratings of buildings in their communities to compare and prioritize relative risks and to form a basis for developing long-term resilience policy. Ultimately, these ratings will benefit our communities by creating market demand for better building construction overall.

Economic and tax incentives

Both positive economic incentives (e.g., subsidies, grants) and negative ones (e.g., fines, penalties) encourage individuals to undertake protective measures. The way that people process information on how these incentives affect the costs and benefits of reducing the risk plays an important role in their decision on whether to adopt the measures (Box 2.8). Several resilience tactics that can reduce business interruption losses after a disaster hits would include those in Box 2.9. Although these resilience tactics are implemented after an event, resilience capacity can be enhanced before an event by actions such as increasing inventories, identifying alternative supply-chain sources and operating locations, and conducting emergency planning drills. In addition, many inherent features of the operation of organizations and the economy as a whole can contribute to resilience. These features can refer to survival motivations that cause businesses and households to make appropriate resilience decisions, as well as the resilience inherent in the market system, where increased prices signal resource shortages and spur more efficient resource allocation (Rose, 2009; NRC, 2011c). Another approach is through tax incentives. For example, if a homeowner or business owner reduces the chances of damage from a hurricane by installing a mitigation measure, the taxpayer could be eligible for a rebate on state taxes to reflect the lower costs for disaster relief.

Hazard forecasting and warning systems

More detailed weather forecasts of the path and severity of a tropical storm or NOAA-developed tsunami warning alerts for U.S. coastal regions can be a key to timely evacuation decisions or movement to shelters (Appendix C). Improvement in the precision of these forecasts is critical for both averting disasters and minimizing their impacts (NRC, 2006a). The broadening of disaster losses to include longer-term impacts and indirect costs such as business interruption (see Chapter 3) has made forecasting more complex. Better and timelier data on the systemic risks also affect the lightly regulated but highly leveraged financial products such as catastrophe bonds.

In addition to forecasting, timely and effective warning about an impending hazard can reduce loss of life and the impacts of disasters by giving people time to take shelter or flee the area (UNISDR, 2007). Early warning systems such as sirens, e-mail, or targeted cell-phone alerts are effective for tornadoes and flash flood events. In all cases, tying early warning systems closely to education and communication programs are critical to develop citizen understanding of the hazard and the actions that citizens can take to plan for and respond to an event (see also Chapter 5; Appendix C).

BOX 2.9
Examples of Post-Disaster Economic Resilience Tactics

- *Conservation*—maintaining production with fewer inputs.
- *Input substitution*—shifting input combinations to achieve the same function or level of productivity.
- *Inventories*—maintaining both emergency stockpiles and ordinary working supplies of production inputs.
- *Excess capacity*—temporarily reducing capacity by idling plant and equipment.
- *Relocation*—changing the site of business activity.
- *Resource independence*—continuing the portion of business operation that can go on without a critical input.
- *Import substitution*—importing resources from other regions, including new contractual arrangements.
- *Technological change*—finding new ways to restore functions, to increase production, to change hours of operation, and to respond to altered product demands.
- *Production recapture*—working overtime or extra shifts to recoup lost production.

Insurance

Insurance is an economic means to allow financial risk to be transferred from a single entity to a pooled group of risks through a contract (Kunreuther and Roth, 1998). The insured party receives an amount of coverage against an uncertain event (e.g., damage to property from an earthquake) in return for a smaller but certain payment (the premium). For a risk to be insurable the insurer has to be able to quantify or estimate the likelihood of the event occurring and the extent of claims when providing different levels of coverage, and to have the ability to set premiums for each potential customer or class of customers.

With respect to flood damage insurers refused to provide coverage since the 1927 Mississippi flood due to their inability to accurately assess the risk. The NFIP was established in 1968 as a result of increased federal relief triggered by disasters in the 1960s and the absence of adequate supply of insurance to cover this hazard (King, 2011). Box 2.10 discusses strategies for modifying the NFIP so that it encourages property owners to undertake mitigation measures in advance of a disaster so that their community is more resilient with respect to flooding.

Catastrophe bonds

To deal with a catastrophic loss, insurers, reinsurers, and governments can use alternative risk transfer instruments such as catastrophe bonds. The use of these financial instruments that take advantage of funds from the capital market grew out of a series of insurance capacity crises in the 1970s through the 1990s that led purchasers of traditional reinsurance coverage to seek more robust ways to buy protection. Catastrophe bonds can enable a country or an organization to access funds from investors if a severe disaster produces large-scale damage in return for premiums the organization pays for a prespecified amount of financial protection (Mahdyiar and Porter, 2005). Suppose the losses to an insurer from an earthquake in California covered by a cat bond exceed a pspecified trigger (e.g., $10 million). Then funds from the cat bond are provided to the insurer to cover a portion of the insurer's claims payments.[10]

BOX 2.10
Making Flood Insurance Work

As of April 30, 2012, the National Flood Insurance Program (NFIP) covers more than $1.26 trillion of property, over five times what was covered 20 years ago.[a] However, people residing in flood-prone areas, many of whom are required to purchase coverage as a condition for obtaining a mortgage, often do not have flood insurance for a variety of reasons. For example, in Vermont, there are only 4,135 flood insurance policies (as of January 2012), which cover 1.3 percent of all housing units in the state.[b] Yet, many property owners residing in areas inundated by Hurricane Irene did not have flood insurance to cover the damage to their homes. Some of them were not able to buy a policy from the federal government because their community did not choose to enroll in the NFIP. Others did not perceive that they would be subject to water damage from hurricanes or tropical storms and chose not to purchase insurance. Still others in floodplains own their homes outright and thus are not subject to the mortgage requirements that otherwise require flood insurance.

Property owners commonly purchase coverage after they suffer losses from a disaster but then cancel their policies several years later if they have not suffered damage again because they view insurance as a poor investment (Kunreuther et al., in press). Difficulties arise in convincing people that they should celebrate their good fortune by recognizing that no return at all on their policy is the best return possible.

Currently, insurance premiums in the NFIP do not necessarily reflect the actual risks faced. This remains a challenge to the program because individuals may, as a result, not recognize the severity of the hazards they face.

[10] For more details on the nature of catastrophe bonds and other alternative risk transfer instruments see Kunreuther and Michel-Kerjan (2011, Chapter 8).

Furthermore, reductions in insurance premiums are not awarded today, even if property owners invest in mitigation measures.

FEMA is now updating its flood maps to more accurately estimate the likelihood and potential consequences of future flooding to property at risk (NRC, 2007a, 2009). Premiums in many areas are likely to be higher than they are today and this increase could have a severe impact on low income and other households that need special treatment. For such reasons, people may not be enthusiastic about flood mapping even though more accurate flood maps can help individuals and communities assess flood risk. In cases where socially vulnerable members of a community may have difficulty paying insurance premiums as a result of new flood maps, insurance vouchers (similar in concept to food stamps) could be an option provided through federal programs. The NFIP was renewed in July 2012, and the legislation suggests that the Federal Emergency Agency and others examine ways to incorporate risk-based premiums coupled with means-tested insurance vouchers.[c]

One way to achieve resilience may also be to tie multiyear insurance policies in flood-prone areas to the property rather than to the individual to avoid cancellation of insurance. Enforcement of building codes through third-party audits by certified building inspectors could also help improve resilience. Home improvement loans for encouraging investment in loss reduction measures could be offered by banks with accompanying reductions in the cost of insurance to reflect the lower risk. In many cases the reduction in annual premiums may be greater than the annual loan payments. In these situations mitigating homes could be viewed as financially attractive. By modifying flood insurance in this way, we may avoid many of the problems faced, for example, by residents in the Northeast following Hurricane Irene (Michel-Kerjan and Kunreuther, 2011).

[a]http://bsa.nfipstat.com/reports/1011.htm.

[b]http://www.fema.gov/library/viewRecord.do?id=4566.

[c]http://www.gpo.gov/fdsys/pkg/BILLS-112hr4348enr/pdf/BILLS-112hr4348enr.pdf.

TABLE 2.2 Illustrative Risk Management Tools, Actors, Time Frames, Benefits, and Potential Adverse Impacts

Illustrative Risk Management Tools	Relevant Individuals, Groups and Organizations)	Time Frame	Potential Benefits	Potential Adverse Impacts
Structural (construction-related)				
Levees, dams, and floodways	USACE, USGS, FEMA; state, county, and local governments; researchers; private sector	1-2 years, evaluation and decision; 3-50 years, construction	Flood risk reduction	Belief that levee will fully protect against all floods
Disaster-resistant construction and retrofitting of existing building stock	Federal government, local officials, researchers, private sector (risk management firms and engineering firms), professional organizations, individuals	On the order of weeks to years depending on measures employed and size of structure	Mitigation against extreme weather events, other natural hazards such as earthquakes and wildfires, floods, and hurricanes	Cost of the measures, belief that energy dissipation systems will fully protect against all hazards
Hazard-conscious ("smart")	Engineering and construction firms individual	Similar to or slightly longer time to build than other new homes	Mitigation against a variety of natural hazards; reduce losses	Cost may be higher than with non-hazard-conscious buildings

building	businesses and homeowners federal and state governments professional organizations	or businesses		
Securing building components and contents	Building owners, tenants	At the time the building is occupied	Reduce earthquake (or other) damage for low investment	None
Well-enforced building codes	Local officials working in conjunction with USGS, NOAA, USFS, FEMA, NIST; engineering firms and professional safety and engineering organizations; businesses and homeowners	Code development: months to several years to review and revise existing codes relative to existing hazard risk Code enforcement: continuous	Home owners and business owners adopt mitigation measures; seals of approval on homes to show that property meets mitigation standards	Inability of some residents to afford compliance and lack of safety net
Nonstructural (nonconstruction-related)				
Natural defenses	Communities, regions, states, federal government	1-4 years: evaluation and decision; from 3 years to many lifetimes	Protect structures built behind them by reducing impacts of disasters (wind, water, fire)	May prevent building new structures on protected areas; requires long-term perspective

Risk mapping	FEMA, USACE, NOAA, NASA, USFS, USGS in conjunction with state and local authorities; engineering firms	Weeks to several years depending upon quality and availability of data and map area covered	Communication of the hazard risk to the community	Overreliance on accuracy of maps
Zoning ordinances	Local and state governments	Immediate	Prohibits building or rebuilding in hazard-prone locations	May prevent lucrative construction of homes or businesses in specific areas
Hazard and vulnerability disclosure	Private sector; federal, state, and local governments	Immediate if adopted freely by the private sector; several years or more if new legislation is required to implement	Allows buyers to identify potential hazards or construction known to be vulnerable to such hazards before the purchase of a home or business; increases the value of disaster-resistant buildings	May hinder sales or lower property values in areas where hazards are revealed or for vulnerable construction types
Economic and tax incentives	Federal, state, and local governments	May be quickly adopted and implemented if political will, competing demands for resources, and public acceptance align; realization of returns on investment	Subsidies, grants, fines, or tax rebates can provide incentives to homeowners and businesses to install hazard mitigation measures	Negative incentives (fines, penalties) may not be acceptable to residents or businesses; positive incentives (subsidies, grants, rebates) incur immediate costs to the government with delayed return on investment

		may be months to many years		
Hazard forecasting and warning systems	NOAA, USGS, USACE, NASA, USFS, state agencies, private sector	Constant data collection and monitoring	Allows forecasts of potential events and their impacts to be made; when communicated in a timely way, warning systems can save lives	Complex disasters and natural systems, increasing population, and potential longer-term impacts require increased data precision and better forecasting models
Insurance	FEMA, state insurance commissioners, private insurance industry, banks	Policies currently are issued on an annual basis but some consideration is being given to multiyear insurance tied to the property	Risk-based pricing that communicates level of risk to people in hazard-prone areas; vouchers for lower-income owners	Continued public financial assistance to those who do not buy insurance
Catastrophe bonds	Insurers, banks, investors	Typically 1 to 3 years	Risk is transferred to a broad investor base in the event of a catastrophic event; allows access to large fund amounts fairly quickly	Investors lose invested funds if a catastrophic event occurs; insurers pay bond amount with interest if the event does not occur

Note: FEMA = Federal Emergency Management Agency, NASA = National Aeronautics and Space Administration, NIST = National Institute of Standards and Technology, NOAA = National Oceanic and Atmospheric Administration, USACE = U.S. Army Corps of Engineers, USFS = U.S. Forest Service, and USGS = U.S. Geological Survey.

IMPROVING RESILIENCE THROUGH RISK MANAGEMENT

Several themes emerge from disaster risk management, which provide a foundation for increasing the resilience of communities to hazard and disaster risks (Sayers et al., 2012):

1. **Risk cannot be eliminated completely, so some residual risk will continue to exist and require management actions.** The impacts of past natural disasters, particularly recent ones, are not necessarily a key to the future for several reasons. Society and its support systems have become increasingly interdependent (Chapter 1). In addition, human activity and development have destroyed much of nature's defenses against natural hazards. This fact, coupled with likely changes in the physical environment due to climate change, suggests that future hazard probability and exposure will rise if no actions are taken. Historic records are short in a geological time frame, and the possibility exists for more severe floods, earthquakes, or other disasters.

2. **The nature of risk perceptions and behavioral biases are important to consider in developing risk management strategies.** The public and decision makers often underestimate the likelihood of a disaster occurring and hence do not undertake risk-reducing measures beforehand. Short-term strategies may also dominate when deciding what action to take. These behavioral features need to be considered when determining what types of risk management strategies are likely to increase resilience to disasters.

3. **A diverse portfolio of disaster risk management measures provides options for decision makers and communities before, during, and after disasters. Such a portfolio can aid in efficient use of resources and more effective risk management.** A portfolio with diverse risk management measures provides multiple options for enhancing resilience to a community in case one of the measures should fail. Combining well-enforced building codes and insurance with structural reinforcements or other measures can take on special significance to protect the community or region against physical and financial losses should structural measures (e.g., dams and levees, natural defenses) fail to provide full protection against the hazard. A key balance is that between investment in resources for managing disaster and the likelihood and magnitude of the hazards.

4. **The need for science-based objective hazard identification and risk assessments is a critical input into the risk management process**. Such input should be easily communicated to the community, with information and data that are transparent and not cloaked in an unpublished model, with all details proprietary. The sole reliance on anecdotal information, past experience, or deterministic scenarios does not provide an adequate or rigorous foundation for determining disaster risk.

5. **Reflecting risk in insurance premiums while keeping insurance acquisition affordable to those requiring special treatment can encourage more individuals to purchase insurance policies.** When insurance premiums are based on risk they provide signals about the hazards individuals face and can encourage them to adopt cost-effective mitigation measures to decrease their vulnerability to future disaster losses. General public funding, as opposed to insurance premium subsidies, can provide insurance for homeowners *currently residing* in hazard-prone areas and who may be socially vulnerable but are uninsured or inadequately insured.

6. **Communicating risk in ways that are understandable to the public is a critical aspect of the risk management process.** Decision makers and the public require accurate information on the risks they face. Risk maps, framing of information, social networking, and educational processes can be employed to communicate information on the risk and on mitigation measures (Sayers et al., 2012; this topic is addressed in detail in Chapter 5).

KNOWLEDGE AND DATA NEEDS

To achieve resilience the federal government has a dominant leadership role in supporting research to improve forecasting, impact-modeling capabilities, as well as the efficacy of risk-reduction strategies for the physical, public health, ecological, and socioeconomic aspects of natural and human-made disasters. Over the last several decades, significant investment by federal and state agencies in both land-based and space-based monitoring and observation networks for natural hazards has greatly increased our ability to forecast the likelihood and characteristics (e.g., magnitude, path) of future event occurrence as well as the intensity of the physical impacts of natural hazard events (e.g., ground-shaking level, wind speed, inundation depth). These data networks provide a quantitative basis for accurate, real-time meteorological forecasting, as well as early warning of flooding and tsunamis. In addition, these hazard monitoring networks provide a multidecadal baseline to help evaluate natural variability as well as the impacts of climate change.

The digital technological revolution made hazard monitoring network data available in real time and, in some cases, permitted rapid computer-automated, preliminary data analysis. The nation relies on a number of essential land-based and space-based hazard monitoring networks for short-term forecasting and early warning, as well as for understanding the physical processes leading to natural disasters and their physical impacts. Both the sensors and the communication networks supporting them require continual maintenance as well as upgrades to take full advantage of technological advances in sensor capabilities and communications. However, resource limitations have prevented many federally run monitoring networks from taking full advantage of the technological advances. The key federal hazard monitoring networks (along with the relevant reviews which include recommendations) are illustrated in Appendix C. Nearly all these networks have been the subject of outside reviews with consistent recommendations for upgrades. While it is beyond the scope of this report to repeat all the recommendations related to hazard monitoring in each of the NRC reports listed in Appendix C, we extend our strongest support for continued and adequate upgrading, expansion of coverage, maintenance, and staffing of the key hazard monitoring networks and observation platforms as outlined above. ***These data are essential for sustaining the forecasting and modeling capabilities required for national resilience.***

Achieving resilience involves monitoring impacts in all the systems and the integration of data. While many hazard monitoring networks are in place, comparable networks for monitoring changes in the human systems as they affect vulnerability and resilience are lacking. Monitoring vulnerability and resilience requires long-term systematic data collection to capture for place-based human and environmental changes. A number of studies have advocated establishing place-based observatory networks on community resilience and vulnerability (Peacock et al., 2008; NRC, 2011c)—observatories that integrate social sciences, natural sciences, and engineering data in monitoring progress toward resilience.

Breakthroughs in hazard and risk assessment will come from better constraints on the key parameters in the models that govern the systems responsible for disaster impacts, such as the role of clouds in climate models, the three-dimensional effects of basins on strong ground shaking in earthquakes, and improved estimates of seasonal and diurnal changes in populations in hazardous areas. Research is also needed on the role and function of natural defenses against natural disasters (e.g., the capacity of coastal wetlands to help absorb storm surge, the role of swamps along rivers for floodwater storage), many of which have been severely compromised by actions of people. Until we fully understand the full ecosystem functions and feedback loops of these natural defenses, it is difficult to meaningfully evaluate whether it would be more cost-effective to restore wetlands or swamps or simply build or continue to raise and strengthen a system of levees downstream.

Research is also scant on the value of disaster mitigation and what factors strongly reduce losses. Targeted research into new materials and new processes for much more resilient construction of new buildings and infrastructure is needed, as well as assessment models of the role of retrofit standards to meet resiliency goals or effective strategies for addressing infrastructure interdependencies, . From a social science perspective, more research is required in modeling social capital within communities. Integration of information and modeling the connections between threats, vulnerability, exposure, sensitivity, and impacts also require more research, especially based on differences in geographic scale or time periods.

One of the key themes in the report is that despite some level of information about disaster risk, individuals, communities, businesses, and political leaders may be reluctant to reduce risk to make the nation more resilient. The question is why? To address that question more research into the social and behavioral biases that affect the processing of risk information, how risk data could be more effectively communicated, and how such risk information translates into the adoption of resilience strategies could be helpful. Research on the next generation of technologies for communicating and sharing location-based risk information would also enhance resilience at all levels.

SUMMARY AND RECOMMENDATIONS

Understanding, managing, and reducing risk is an essential foundation for increasing resilience to hazards and disasters. Risk management is a continuous process, and the choice of strategies requires regular reevaluation in the context of new data, models, and changes in the socioeconomic and demographic characteristics, and environmental setting of a community. The risk management strategy that works best for a community is based on the available information, how it is communicated to the key interested parties, and the perception of risks and rewards for avoiding or mitigating risk.

A variety of tools exists to manage disaster risk. These tools include structural (construction-related) measures such as levees, dams, disaster-resistant construction, and well-enforced building codes, and nonstructural (nonconstruction-related) measures such as natural defenses, insurance, zoning ordinances, and economic incentives. Structural and nonstructural measures are complementary and can be used in conjunction with one another. Risk management is at its foundation a community decision—including not only the immediately affected community, but also local, state, and federal levels of government and the private sector—and the risk management approach and will only be as effective if there is commitment to use risk management tools and measures.

Recommendation: **The public and private sectors in a community should work cooperatively to encourage commitment to and investment in a risk management strategy that includes complementary structural and nonstructural risk-reduction and risk-spreading measures or tools.** The portfolio of tools should seek equitable balance among the needs and circumstances of individuals, businesses, and government, as well as the community's economic, social, and environmental resources.

Examples from actual disasters and their aftermaths show that implementation of risk management strategies involves a combination of actors in local, state, and federal governments, NGOs, researchers, the private sector, and individuals in the neighborhood community. Each actor will have different roles and responsibilities in developing the risk management strategy and in characterizing and implementing the measure or tool, whether structural or nonstructural, to be added to the community's risk management portfolio.

Some strategies can be implemented over the short term, while others may take a longer time. Table 2.2 is a potential template for decision makers to consider how to develop and implement risk management strategies and to manage expectations. The roles and responsibilities of the different actors are described in more detail in Chapters 5 and 6.

One underutilized tool is investment in risk reduction through insurance and other financial instruments to enhance resilience. Such measures can improve mitigation of properties and infrastructure, but more importantly, can encourage the relocation of residences, businesses, and infrastructure through more risk-based pricing.

Recommendation: **The public and private sectors should encourage investment in risk-based pricing of insurance in which insurance premiums are designed to include multiyear policies tied to the property, with premiums reflecting risk.** Such risk-based pricing reduces the need for public subsidies of disaster insurance. Risk-based pricing can serve as an incentive that clearly communicates to those in hazard-prone areas the different levels of risk that they face. Use of risk-based pricing could also reward mitigation through premium reductions and can apply to both privately and publicly funded insurance programs.

REFERENCES

Bachman, R. E. 2004. The ATC 58 project plan for nonstructural components In Proceedings of the International Workshop on Performance-Based Seismic Design, Concepts and Implementation, June 28-July 1, 2004, Bled, Slovenia, P. Fafjar, ed. Richmond, CA: Pacific Earthquake Engineering Research Center.

Bradley, A. A., Jr. 2010. What causes floods in Iowa? In A Watershed Year: Anatomy of the Iowa Floods of 2008, C. F. Mutel (ed.). Iowa City: University of Iowa Press, pp. 7-18.

Brinkley, D. 2006. The Great Deluge: Hurricane Katrina, New Orleans, and the Mississippi Gulf Coast. New York: Harper Collins.

Burby, R. J. 2006. Hurricane Katrina and the paradoxes of government disaster policy: Bringing about wise governmental decisions for hazardous areas. Annals of the American Academy of Political and Social Science 604:171-191.

Camerer, C., and H. Kunreuther. 1989. Decision processes for low probability events: Policy implications. Journal of Policy Analysis and Management 8:565-592.

Cornell, C. A. 1968. Engineering seismic risk analysis. Bulletin of the Seismological Society of America 58:1583–1606.

Cutter, S. L., ed. 2001. American Hazardscapes: The Regionalization of Hazards and Disasters. Washington, DC: Joseph Henry Press.

Cutter, S. L., B. J. Boruff, and W. L. Shirley. 2003. Social vulnerability to environmental hazards. Social Science Quarterly 84 (1):242-261.

Cutter, S. L., L. Barnes, M. Berry, C. Burton, E. Evans, E. Tate, and J. Webb. 2008. A place-based model for understanding community resilience to natural disasters. Global Environmental Change 18(4):598-606.

Cutter, S., B. Osman-Elasha, J. Campbell, S.-M. Cheong, S. McCormick, R. Pulwarty, S. Supratid, and G. Ziervogel. 2012. Managing the risks from climate extremes at the local level. In Managing the Risks of Extreme Events and Disasters to Advance Climate Change Adaptation. A Special Report of Working Groups I and II of the Intergovernmental Panel on Climate Change, C. B. Field, V. Barros, T. F. Stocker, D. Qin, D. J. Dokken, K. L. Ebi, M. D. Mastrandrea, K. J. Mach, G.-K. Plattner, S. K. Allen, M. Tignor, and P. M. Midgley, eds. Cambridge, UK, and New York: Cambridge University Press, pp. 291-338.

Elnashai, A. S., V. Papanikolaou, and D. Lee. 2009. ZEUS-NL—A system for inelastic analysis of structures. Mid-America Earthquake Center, Univ. of Illinois at Urbana-Champaign, IL. □http://mae.cee.uiuc.edu/news/zeusnl.html□.

Emrich, C. T., S. L. Cutter, and P. J. Weschler. 2011. GIS and emergency management. In The SAGE Handbook of GIS and Society. Thousand Oaks, CA: SAGE, pp. 321-343.

FEMA (Federal Emergency Management Agency). 1998. Retrofitting: Six Ways to Prevent Your Home from Flooding. Washington, DC: FEMA.

Galloway, G., G. Baecher, D. Plasencia, K. Coulton, J. Louthain, and M. Bagha. 2006. Assessing the Adequacy of the National Flood Insurance Program's 1 Percent Flood Standard. Washington, DC: American Institutes for Research.

Galloway, G. E, D. F. Boesch, and R. R. Twilley. 2009. Restoring and protecting coastal Louisiana. Issues in Science and Technology 25:2:29-38.

Grossi, P., and H. Kunreuther. 2005. Catastrophe Modeling: A New Approach to Managing Risk. New York: Springer.

Heinz Center. 2000. The Hidden Costs of Coastal Hazards. Washington, DC: Island Press.

Huber, O., R. Wider, and O. Huber. 1997. Active information search and complete information presentation in naturalistic risky decision tasks. Acta Psychologica 95:15-29.

IAEM (International Association of Emergency Managers). 2007. Principles of Emergency Management Monograph (Supplement). Available at https://www.iaem.com/EMPrinciples/index.htm.

International Bank for Reconstruction and Development/World Bank. 2010. Natural Hazards. UnNatural Disastes: The Economics of Effective Prevention. Washington, DC: World Bank.

IPCC (Intergovernmental Panel on Climate Change). 2012. Managing the Risks of Extreme Events and Disasters to Advance Climate Change Adaptation, C. B. Field, V. Barros, T. F.

Stocker, D. Qin, D. J. Dokken, K. L. Ebi, M. D. Mastrandrea, K. J. Mach, G.-K. Plattner, S. K. Allen, M. Tignor, and P. M. Midgley, eds. Cambridge, UK and New York: Cambridge University Press.

Kahneman, D., and A. Tversky. 2000. Choices, Values and Frames. New York: Cambridge University Press.

Karl, T. R., J. M. Melillo, and T. C. Peterson. 2009. Global Climate Change Impacts in the United States. New York: Cambridge University Press.

Keeney, R. 1992. Value Focused Thinking: A Path to Creative Decisionmaking. Cambridge, MA: Harvard University Press.

King, R. O. 2011. National Flood Insurance Program: Background, Challenges, and Financial Status. Congressional Research Service. Available at: http://www.fas.org/sgp/crs/misc/R40650.pdf.

Krajewski, W. F., and R. Mantilla. 2010. Why were the 2008 floods so large? In A Watershed Year: Anatomy of the Iowa Floods of 2008. Iowa City: University of Iowa Press, pp. 19-30.

Kunreuther, H. 1996. Mitigating disaster losses through insurance, Journal of Risk and Uncertainty 12:171-187.

Kunreuther, H., and E. Michel-Kerjan. 2011. At War with the Weather: Managing Large-Scale Risks in a New Era of Catastrophes. Cambridge, MA: MIT Press (paperback ed.).

Kunreuther, H., and R. Roth, Sr., eds. 1998. Paying the Price: The Status and Role of Insurance Against Natural Disasters in the United States. Washington, DC: Joseph Henry Press.

Kunreuther, H., N. Novemsky, and D. Kahneman. 2001. Making low probabilities useful. Journal of Risk and Uncertainty 23:103-120.

Kunreuther, H., R. Meyer, and E. Michel-Kerjan. 2012. Overcoming decision biases to reduce losses from natural catastrophes. In Behavioral Foundations of Policy. Princeton, NJ: Princeton University Press.

Kunreuther, H., M. Pauly, and S. McMorrow. In press. Insurance and Behavioral Economics: Improving Decisions in the Most Misunderstood Industry. New York: Cambridge University Press.

Mahdyiar, M., and B. Porter. 2005. The risk assessment process: The role of catastrophe modeling in dealing with natural hazards. In Catastrophe Modeling: A New Approach to Managing Risk. New York: Springer, pp. 45-68.

Magat, W., K. W. Viscusi, and J. Huber. 1987. Risk-dollar tradeoffs, risk perceptions, and consumer behavior. In Learning About Risk. Cambridge, MA: Harvard University Press, pp. 83-97.

Michel-Kerjan, E., and H. Kunreuther. 2011. Redesigning flood insurance. Science 333(6041):408-409.

Milly, P. C. D., J. Betancourt, M. Falkenmark, R. M. Hirsch, Z. W. Kundzewicz, D. P. Lettenmaier, and R. J. Stouffer. 2008. Stationarity is dead—Whither water management? Science 319:573-574.

NRC (National Research Council). 2006a. Completing the Forecast: Characterizing and Communicating Uncertainty for Better Decisions Using Weather and Climate Forecasts. Washington DC: The National Academies Press.

NRC. 2006b. Facing Hazards and Disasters: Understanding Human Dimension. Washington DC: The National Academies Press.

NRC. 2007a. Elevation Data for Floodplain Mapping. Washington DC: The National Academies Press.

NRC. 2007b. Successful Response Starts with a Map: Improving Geospatial Support for Disaster Management. Washington DC: The National Academies Press.

NRC. 2009. Mapping the Zone: Improving Flood Map Accuracy. Washington DC: The National Academies Press.

NRC. 2010. Risk in DHS: Review of the Department of Homeland Security's Approach to Risk Analysis. Washington DC: The National Academies Press.

NRC. 2011a. America's Climate Choices. Washington DC: The National Academies Press.

NRC. 2011b. Increasing National Resilience to Hazards and Disasters: The Perspective from the Gulf Coast of Louisiana and Mississippi, Summary of a Workshop. Washington, DC: The National Academies Press.

NRC. 2011c. National Earthquake Resilience: Research, Implementation, and Outreach. Washington, DC: The National Academies Press.

NRC. 2012. Dam and Levee Safety and Community Resilience: A Vision for Future Practice. Washignton, DC: The National Academies Press.

Opperman, J. J., G. E. Galloway, J. Fargione, J. F. Mount, B. D. Richter, and S. Secchi. 2009. Sustainable floodplains through large-scale reconnection to rivers. Science. 326:1487-1488.

Peacock, W. G., H. Kunreuther, W. H. Hooke, S. L. Cutter, S. E. Change, and P. R. Berke. 2008. Toward a Resiliency and Vulnerability Observatory Network: RAVON. Final Report. Hazard Reduction and Recovery Center, Texas A&M University. Available at http://www.nehrp.gov/pdf/RAVON.pdf.

Peduzzi, P., H. Dao, C. Herold, and F. Mouton. 2009. Assessing global exposure and vulnerability to hazards: The Disaster Risk Index. Natural Hazards Earth System Science 9:1149-1159.

Rose, A. 2009. Economic Resilience to Disasters: CARRI Research Report 8. Community and Regional Resilience Initiative, Oak Ridge, TN. Available at http://www.resilientus.org/library/Research_Report_8_Rose_1258138606.pdf.

Samuelson, W., and R. Zeckhauser. 1988. Status quo bias in decision making. Journal of Risk and Uncertainty 1:7-59.

Sayers, P., G. Galloway, E. Penning-Rowsell, F. Shen, K. Wang, Y. Chen, and T. Le Quesne. 2012. Flood Risk Management: A Strategic Approach—Consultation Draft. UNESCO on behalf of WWF-UK/China and the General Institute of Water Design and Planning, China.

Slovic, P. 1987. Perception of risk. Science 236:280-285.

Slovic, P. 2000. The Perception of Risk. Sterling, VA: Earthscan.

Slovic, P., B. Fischhoff, and S. Lichtenstein. 1978. Accident probabilities and seat belt usage: A psychological perspective. Accident Analysis and Prevention 10:281-285.

Tobin, G. A. 1995. The levee love affair: A stormy relationship? Water Resources Bulletin 31(3):359-367.

UNISDR (United Nations International Strategy for Disaster Reduction Secretariat). 2007. Hyogo Framework for Action 2005-2015: Building the Resilience of Nations and Communities to Disasters. Geneva, Switzerland: UN/ISDR. Available at http://www.unisdr.org/files/1037_hyogoframeworkforactionenglish.pdf

"We lost 31 of those (street) cars. To rebuild those cars cost us $1.2 million per car. That's not a capital cost you can replace very easily."
--Justin Augustine, CEO of the New Orleans Regional Transit Authority, January 2011 on losses to the New Orleans transportation system after Hurricane Katrina

3

Making the Case for Resilience Investments: The Scope of the Challenge

INTRODUCTION

The potential benefits of being resilient to hazards and disasters make abundant sense. Few would oppose taking action to reduce the loss of life or property damage. However, increasing the resilience of a community requires large-scale investments of money, human resources, and time. Once risk has been identified and assessed, what actions are sufficient to address the risk? How resilient does the community need to be? How do investments in improving resilience compete with other community investment priorities? What are the benefits? Who pays now? Who pays later?

The available data portraying past disasters show that the scale and scope of disaster losses[1] are enormous and that significant investment is required to mitigate the losses of human life, risks to human health, and economic and social costs. Investments are required for a wide spectrum of community needs such as planning, organizing, training, and equipping first responders to large infrastructure projects. Owners of community assets are primarily responsible for their own resilience investments, yet community leaders from both the public and private sectors recognize that community assets are interconnected and interdependent and that holistic planning, programming, investing, and execution create common and interrelated resilience benefits for the community. Realizing the maximum benefits requires close collaboration among public- and private-sector leaders aided by a shared approach and commitment to investment.

---

[1] Unless otherwise noted, economic losses refer to property damage or crop losses (or both, if noted).

As stewards of community assets the potential benefits of being resilient to hazards and disasters are attractive from governmental, economic, social, and environmental points of view. Although consensus generally exists on the goals for strengthening resilience, making the case for investing in resilience programs, in individual initiatives or projects, and in strengthening weak infrastructure is very challenging, especially in the context of demand for competing resources. Particularly during times of economic hardship, competing demand for many societally relevant resources (education, health, and social services) can be a major barrier to making progress in building resilience in communities. As a prerequisite for making the case, advocates are required to demonstrate that the potential benefits of being resilient to hazards and disasters make conceptual sense. However, such efforts also have to show clearly that community investments in resilience will yield significant and measurable short- and long-term benefits that balance or exceed the costs. This kind of cost-benefit analysis is critical for sustained commitment to increasing resilience, given the rising level of competition for scarce resources at local, state, and federal levels (Rose et al., 2007).

Furthermore, increasing resilience is tied in important ways to economic recovery after a disaster. Specifically, resilience measures can encourage efficient use of existing resources, and thereby lead to as rapid a recovery as possible. Some factors that have been shown to have achieved these ends include rapid business relocation (because of the existence of excess office space), use of inventories and stockpiles, and substitution of inputs or suppliers (Rose and Blomberg, 2010).

One approach that communities can use as they embark on a process of improving resilience is to develop multiyear plans or programs that include compelling initiatives or projects. These projects may include improving weak or underfunded community infrastructure such as schools, clinics, and hospitals, and the services which constitute any community. Involving and empowering individuals and families in developing these programs are important because of the ultimate need for individuals to take a share of responsibility in building resilience. Beyond the essential cost-benefit analysis, the value of each initiative or project also rests on the basis of its life-safety, economic, social, public health, and environmental significance. This kind of valuation can assist community leaders with prioritizing investments, decision making, and developing a schedule for implementing their resilience-building strategies.

Resilience investments challenge traditional approaches to "cost-benefit" analysis because communities have many different kinds of assets which are valued differently. Communities have very-high-value assets that are "essential" to keep operating—for example, hospitals, power plants, water and sewage plants, and transportation and communication networks—that usually have a tangible dollar value attached to them, and the costs of disruptions in these services can usually be directly calculated. The social, cultural, and environmental assets of a community also have high "value" but the value is described in cultural and life-quality terms and is more difficult to quantify in financial terms. Such assets include museums, natural landscapes or areas, protected environmental zones, historical buildings, and a health infrastructure that supports prevention and health maintenance throughout the population. Thus the total value of a community's assets—both the high-value structural assets and those with high social, cultural, and/or environmental value—necessitates qualitative and quantitative inputs into a decision-making framework for disaster resilience. Such decision making is going to be difficult for community leaders as they try to address the value of multiple community assets in economic, social, cultural, and environmental terms. Access to reliable data is vital in order to support these kinds of decisions. This chapter identifies the data needed and an approach for valuing assets, planning, programming, and investment decision making for resilience. Specifically, the chapter addresses (1) the challenge of decision making for community leaders in developing their priorities in the context of their risk management findings and conclusions (see also Chapter 2); and (2) the scale and scope of the threat and potential losses from disasters. The ways in which communities might be able to develop or adapt measures of their progress toward resilience are developed in Chapter 4.

CHALLENGE OF RESILIENCE DECISION MAKING FOR COMMUNITY LEADERS

High-value assets of a community are those for which continued operation is essential and urgent for the entire community (e.g., water and power utilities, fuel systems, transportation facilities and systems, communication systems, first responder operations centers, and hospitals). These interdependent, high-value assets drive the need for holistic thinking, risk management (Chapter 2), priority setting, and investment timing.

Although substantial investments in some communities are made for contingency preparations to secure essential community services and operations during disasters, the scale of a disaster can nonetheless overwhelm the capacity of the system and its operators to cope, leading to a failure in one or more parts of the system as occurred, for example, with essential utilities in coastal Louisiana during and after hurricane Katrina (NRC, 2011). Proven techniques such as systemwide analyses and scenario planning offer insights for decision makers to see resilience improvement needs and weigh their investment priorities.

Other high-value assets of a community may include its economic foundation (e.g., local industry or business), and its social, cultural, environmental, and educational assets. These may include traditional ethnic neighborhoods, religious centers, parks and preserves, wildlife habitats, art centers and architectural icons, town squares, and schools or other educational institutions. These assets are held dear and are highly valued as distinguishing attributes by the community. Although it is difficult to measure their value in purely monetary terms, their loss may significantly degrade the total ambiance or qualify of life of a community. Although such losses may at first be devastating, the investment priority judgments of community leaders will consider the promise and possibilities embedded in the ingenuity and self-reliance of citizens (see Box 3.1).

Establishing ownership of a community's assets is also important. Asset owners in a community will vary and include those from public utilities, local businesses and industries, faith-based communities, governmental and nongovernmental organizations, and individual citizens. Owners are primarily responsible for their property and for making appropriate steps including investments in mitigation measures—structural and nonstructural (see Chapter 2)—to prepare and plan for hazards and risks. Community resilience planning and investment programming set goals, strategies, and metrics for the community and guide owners in how best to prioritize and time their investments. However, resilience is also the outcome of interconnected systems (Chapter 1). Decisions about the prioritization and the level of investment require consideration of both quantitative data and qualitative value assessments

BOX 3.1
Decentralization of Community Assets: One Means to Forge a Greater Sense of Community Resilience

Prior to Hurricane Katrina, the public school system in New Orleans was centralized, and the schools were operated largely through a unified school district and primarily served one community function—to educate the city's children. With the destruction of many essential functions including the schools and school system in New Orleans as a result of Katrina, some members of the private sector, nonprofit organizations, and local citizens revisited together the "value" of their schools in the context of the larger neighborhood communities that the schools serve. What emerged was a design for new schools that encompassed a "systems" approach where schools were designed and built to serve multiple community purposes—with facilities to support cultural and social events and community health through fitness centers in gymnasiums. Investments in hardening the school structures to withstand the hazards present in the area have focused not only on protecting students in the event of a disaster, but also on having the schools capable of serving as centers for shelter of the neighborhood community in case of a crisis. These planned investments by the "owners" and stakeholders of this educational community asset—essentially a blend of private, nonprofit, and community members—have increased the scope of the asset as well as its overall community value.

Source: NRC (2011); Steven Bingler, personal communication, January 20, 2011.

the community is key in this regard. The next section examines the urgency of the need to consider the scale and scope of disasters and disaster losses as a means to motivate community efforts to identify and prioritize the full extent of a community's assets.

THE SCALE AND SCOPE OF DISASTERS AND DISASTER LOSSES— AN URGENT PROBLEM

The Economic Value of Mitigation

Understanding the benefits of investing in one or more mitigation strategies in one place may provide some level of guidance that similar measures implemented elsewhere may yield a certain, or potentially greater, level of benefit. One of the landmark studies on the economic value of disaster mitigation is the work of the Multihazard Mitigation Council (2005), a public–private partnership established to reduce the economic and social costs of natural hazards. The study, based on cost-benefit analysis, examined future savings from hazard mitigation related to earthquakes, wind, and floods using two approaches: (1) a purposive sample of communities with mitigation grants funded by the Federal Emergency Management Agency (FEMA) to determine losses avoided through reductions in direct property damage, business interruptions, nonmarket damages, human losses, and costs of emergency response; (2) estimates of future savings from FEMA mitigation expenditures that use a statistically representative sample of FEMA-funded mitigation grants and that was then generalized to all FEMA mitigation grants (Multihazard Mitigation Council, 2005). HAZUS-MH was used to estimate direct property damage from earthquake, flooding, and hurricane wind. The mitigation approaches included both physical measures (elevating or relocating structures, strengthening structures against earthquake or wind hazards) and processes (such as building codes, policies, education). The study results concluded that mitigation saves money with benefits that greatly exceed the costs: for every $1 spent on pre-event mitigation, $4 was saved in post-event damages (see also Chapter 1). In another study that examined physical mitigation measures, Sutter et al. (2009) found that wind-resistant construction costing less than $500 additional per typical single-family home could mitigate future wind damage in tornado-prone regions by 30 percent. Research conducted by the Institute for Business and Home Safety has also demonstrated the economic value of relatively simple and inexpensive home fortification through significant reduction in structural damage and economic losses from strong weather events (Box 3.2).

BOX 3.2
For the Want of a Ring-Shank Nail, the Roof Was Lost: Research Supports Inexpensive Ways to Fortify a Home against Natural Hazards

Steps toward resilience need not be expensive. During a wind, water, or fire event, the roof is often involved, and "once the roof cover is compromised, all sorts of bad things can happen to the structure" (J. Rochman, personal communication, January 20, 2011). Research conducted by the Institute for Business and Home Safety (IBHS) has demonstrated that using ring-shank nails with full round heads instead of smooth-shank nails or staples to hold siding and roofing materials on a home contributes to significantly more resilient structures when the homes are subjected to strong weather events such as hurricanes and wind storms. IBHS has a stronger, safer construction standard for new homes, known as FORTIFIED for Safer Living®, which goes above building codes (where they exist) with risk-specific guidance for homeowners, architects, and builders.

A simple and inexpensive change to improve the resilience of a roof—whether first put on a new building or during reroofing—is to use a minimum of 2⅜-inch ring-shank nails instead of smooth-shank nails or staples to secure the roof decking. In a series of full-scale tests at the IBHS Research Center, two virtually identical two-story, 1,300-ft^2 homes (one built to standard building codes as they exist in the center of the country and one built to FORTIFIED standards for new construction) were subjected to hurricane-strength wind speeds. Unlike the conventionally constructed house, the FORTIFIED house had ring-shank nails securing the roof and met other FORTIFIED requirements, such as using metal strapping to hold load-path elements together. The cost of the extra FORTIFIED modifications totaled only about $3,000. After subjecting both houses to sustained wind and gusts that peaked at 96 miles per hour, professional insurance adjustors examined both homes and estimated that the cost of exterior repairs to the conventionally built home was ~2.5 to 8 times higher than the home built to the IBHS FORTIFIED standard.

FORTIFIED program value was clearly demonstrated in a real-world situation on the Bolivar Peninsula of Texas during Hurricane Ike. Thirteen FORTIFIED homes stood directly in the path of Ike's eye wall, which included 110-mph winds and an 18-ft to 20-ft storm surge. Ten FORTIFIED homes remained standing with minimal damage, while all other homes for miles around were totally destroyed. The three FORTIFIED homes that were destroyed were lost due to impacts from surrounding homes that were knocked off their foundations and became moving piles of debris.

Research by the committee at a local home supply store revealed the cost of 2,500 2⅜-inch ring-shank nails with full round heads was $38. Approximately 6,000 nails are required to attach the roof sheeting for a 2,000-ft^2 house, another 6,000 nails with plastic or metal caps to anchor the underlayment, and about 12,000 nails to attach the shingles (ca. 6 nails per shingle).

Sources: http://www.disastersafety.org/content/data/file/FORTIFIED-vs-Conventional.pdf; http://www.disastersafety.org/fortified; J. Rochman, personal communication, January 20, 2011.

Patterns of Disaster Losses to Guide Resilience Investments

Examining historic patterns of disaster losses provides some sense of the magnitude of the need to become more disaster resilient. In addition, the geographic patterns of disaster losses—human fatalities, property losses, and crop losses—illustrate where the impacts are the greatest, and where there could be challenges in responding to and recovering from disasters. Geographic patterns of losses, when compared with available data on housing, population growth, income level, and types of natural hazards, allow understanding of some of the driving factors of exposure and vulnerability to hazards and disasters (see also Chapter 2), and can lead more readily to appropriate paths to increase resilience. This kind of analysis also reveals gaps in our knowledge of natural, built, and socioeconomic systems—including their interrelationships—and is useful in prioritizing research needs. The following sections review disaster losses in terms of U.S. and global tendencies; geographic variation in economic losses, human losses, and patterns of exposure; and population growth. Each section draws upon available data and also presents evidence for gaps in data collection, analysis, and availability.

U.S. and Global Patterns in Economic Losses

Because local and national patterns in disaster losses occur within a larger global context, a useful way to assess the current state of resilience in the United States is to examine the magnitude of global events and losses. As estimated by Munich RE (2012), the costliest year on record for natural disasters around the world (based on preliminary global data for the year) was 2011, with more than $380 billion in losses (of which only $105 billion was insured), exceeding the previous record set in 2005. The earthquakes in New Zealand, the March earthquake and tsunami in Japan, and flooding in Australia and Thailand all contributed to these new levels of loss. The Japanese earthquake and tsunami combined were the most costly events globally in 2011. In the United States, estimated losses were $64 billion, of which $35.8 billion were insured losses (Munich RE, 2012). The snows of February, severe storms in April and May which brought many tornadoes, the extensive flooding in the Midwest and Great Plains, wildfires in Texas and the Southwest, and Hurricane Irene impacting much of the U.S. East Coast all contributed to the total (see also Figure 1.1).

Establishing the tendencies in economic losses provides the baseline against which we can monitor losses avoided due to improved resilience. Data that have been collected in a consistent manner are essential for measuring losses in absolute terms over time or in different locations, or simply attempting to monitor loss history for a single location. Existing global loss databases are useful for certain kinds of analyses, but require improvement in measurements, accuracy, and consistency. For example, there is an ongoing debate in the literature over whether losses from natural disasters are actually increasing over time (Figure 3.1), or whether the data reflect large, recent singular extreme events (e.g., the Tohoku earthquake and tsunami), changes in asset values, changes in reporting, changes in housing stock, improved awareness, or some combination of these. When national losses are normalized for population and wealth, upward patterns in normalized losses appear to become less significant (Pielke and Landsea, 1998; Brooks and Doswell, 2001; Miller et al., 2008); however, other evidence suggests that even with normalization for population and wealth, losses are increasing significantly, especially in the United States (Gall et al., 2012) (Figure 3.2). Improvements in disaster-data collection will help clarify these fundamental tendencies.

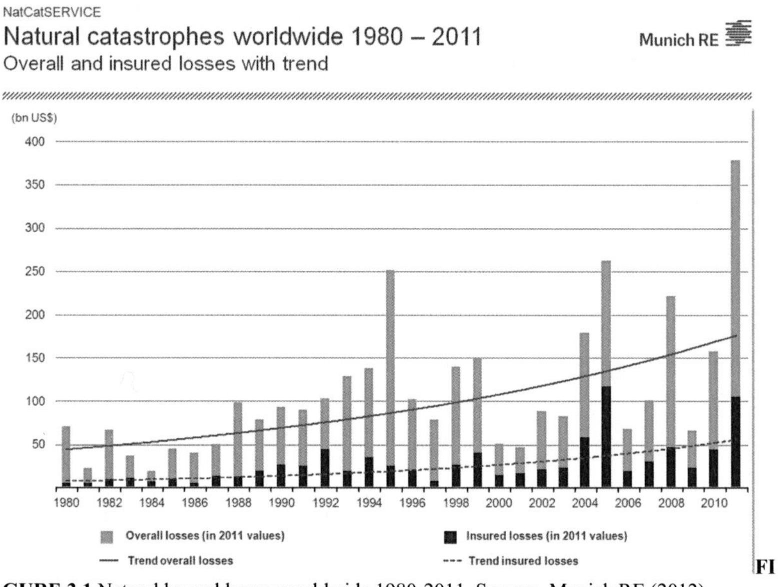

FIGURE 3.1 Natural hazard losses worldwide 1980-2011. Source: Munich RE (2012).

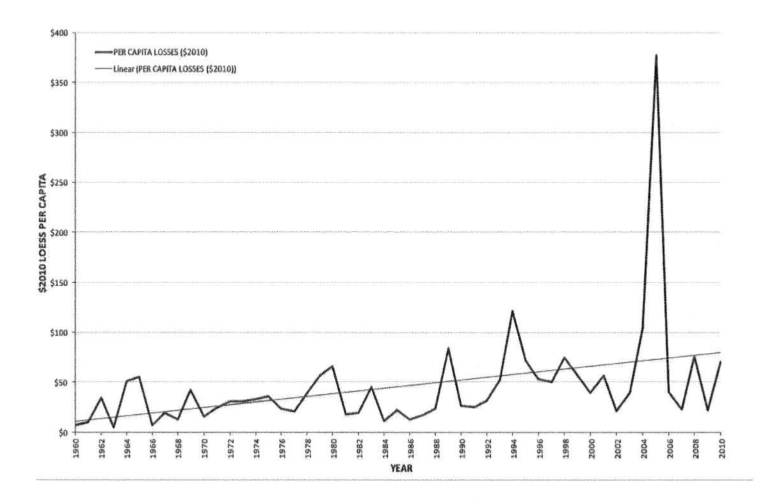

FIGURE 3.2 Trends in per-capita property and crop losses (adjusted to $2010) from natural hazards, 1960-2010. According to Gall et al. (2012), per-capita losses appear to be escalating in the United States, even when normalized by population, and have more than tripled per person since the 1960s. Source: S. Cutter; compiled from SHELDUS.

Another issue in analyzing loss data is that not all losses are counted and valued (Box 3.3). In the case of Munich RE, the NatCatSERVICE database provides property losses (total and insured) and insured business interruption losses, estimated from known insured losses. Because of the differences in loss estimation techniques, thresholds for inclusion in the database (large versus small events; insured versus uninsured losses), and data availability (public versus proprietary), natural-hazard loss databases are rarely comparable with one another. For example, comparisons among four publicly accessible databases show different total dollar loss estimates for the United States in 2010 attributed to weather perils such as winter storms, hurricanes, tornadoes, and flooding (Table 3.1). In the health arena, some losses of life and health may occur days or months after the disaster and thus may go uncounted.

Geographic Variation in Economic Losses

Long-term disaster loss data for specific geographic regions provide a baseline from which to measure improvements in resilience. The success of measures to reduce disaster risk and impacts are difficult to evaluate without this baseline information. A number of federal agencies compile separate data on disaster losses and costs including the National Oceanic and Atmospheric Administration (NOAA), FEMA, the U.S. Geological Survey (USGS), and the Department of Agriculture. These data serve quite specific and useful purposes, but in aggregate are incomplete, often incompatible with one another, have limited economic impact information, and are less useful for mapping the

BOX 3.3
Which Economic Losses Are Counted?

Losses from natural hazards are normally divided into two major categories—direct losses and indirect losses. Economic losses are classified as stock losses (property damage) and flow losses (business interruption). There are direct and indirect versions of each. For example, direct property damage occurs from the seismic shaking from an earthquake whereas indirect property damage can occur from fires due to the rupture of a natural gas pipeline caused by the earthquake. Direct flow losses occur to those businesses in the affected area that had to shut down temporarily. Indirect flow losses refer to the disruption in the supply chain for other businesses as a result of the shutdown (a ripple effect caused by the interconnectedness of many supply chains regionally and globally). Other primary losses include the costs of repair and placement of structures, the cost of debris removal, loss of jobs, loss of rental income, and evacuation costs. Secondary losses such as those associated with decreased tax revenues, decline in property values, loss of attractiveness of tourist destinations, psychological trauma, and the damage to natural systems are not taken into account in loss tallies, yet these hidden costs may directly influence the affected community's ability to manage disaster risk.

SOURCES: Heinz Center (1999), Rose (2004), Multihazard Mitigation Council (2005), NRC (2006a); Gall et al. (2009).

geographic distribution and impact of such losses at the local (community to state) scale. Currently, no comprehensive federal database or national archive for disaster loss data exists (Mileti, 1999; NRC, 1999; Cutter, 2001). The SHELDUS® (Spatial Hazard Event and Loss Database for the United States), compiled from existing federal data sources, is the closest approximation to a U.S. national inventory of direct disaster losses from natural hazards, but it also underestimates the total value of losses because indirect losses and business interruption are not included, for example. Such indirect losses can be substantial (see Box 3.4).

SHELDUS information can be used to examine patterns losses from natural hazards within the United States over the last 50 years. Figure 3.3 shows that these losses tend to be concentrated in a few regions and within a few states. The overall patterns highlight losses on the hurricane coast along the Gulf of Mexico extending from Texas to Florida and up the Atlantic Coast to the Carolinas. When normalized to losses per square mile (Figure 3.3b) the largest cumulative losses are concentrated in California, western Washington, the Gulf Coast and Florida, the Carolinas, the Northeast, and in the upper Midwest.

Table 3.1 Losses from Selected Weather-Related Hazards in the United States for 2010.

Database[a]	Loss ($ Billion)	Deaths
Munich RE	13.6	197
NCDC Billion Dollar Events	6.8	46
SHELDUS	8.8	266
EM-DAT	9.15	90

[a]Munich RE = NatCatSERVICE (which includes total property loss, known insured property losses, and estimated insured business interruption losses; NCDC Billion Dollar Events (http://www.ncdc.noaa.gov/oa/reports/billionz.html#narrative) reported total property and crop loss; SHELDUS = Spatial Hazard Events and Loss Database for the United States, maintained by the Hazards and Vulnerability Research Institute at the University of South Carolina (http://www.sheldus.org), reported total property and crop loss; EM-DAT = Emergency Events Database, maintained by the Centre for Research on the Epidemiology of Disasters, CRED (http://www.emdat.be), estimated property and crop loss, loss of revenues). See Gall et al. 2009 for more details on the databases.

BOX 3.4
Spatial Hazard Event and Loss Database for the United States (SHELDUS®)

SHELDUS is a county-level database for U.S. states of loss-causing natural hazards that spans the period from 1960 to the present. The database is maintained by the University of South Carolina's Hazards & Vulnerability Research Institute. It reports only direct losses as defined by the federal source data it uses (e.g., National Climatic Data Center's Storm Data; U.S. Geological Survey Open File Reports), and does not include Puerto Rico, Guam, or other U.S. territories. The historic Storm Data (1960-1995) used logarithmic categories for losses; for example, an event with a loss category 5 represents losses of $50,000-$500,000 in that database. SHELDUS uses the lower-bound value (e.g., $50,000), and as a result, the database is conservative and provides the minimum value of losses over the specified time period. Thus, losses are expected to be higher than those reported in the database, but how much higher is presently unknown.

The database is available online (http:www.sheldus.org), can be queried by individual hazard, by geography (state and county), by time period, by presidential disaster declarations number, by major named disasters (e.g., Hurricane Katrina, Blizzard of 1967), and by GLIDE number (an international standard numeric to enable linkages across databases). The database provides property losses (recorded in period dollars); crop losses (recorded in period dollars); injuries; fatalities; county, state, and federal Information Processing Standard codes; and beginning and ending dates for when the information was recorded. Losses can be converted to current dollars or standardized to any year by the user. All the data from the queries is downloadable into a spreadsheet. At present, the database (v.10.0) contains over 710,000 records. The strengths of the SHELDUS database relate to its county-level coverage for a 50-year time period for 18 different hazard types. The consistent georeferencing over time, despite changes in county boundaries, is another added feature. The weaknesses of SHELDUS relate to the input data, culled from federal sources. The federal databases were developed for a different purpose; inconsistencies and biases in those data are transferred to SHELDUS. For example, in many reports of weather-related losses, an entire state was given in the record and the database disaggregation technique is to apportion the losses equally across affected counties when no additional data were provided. This technique results in a geographic pattern that may appear state-centric, but in reality is a function of the initial reporting of losses. SHELDUS is a database and does not predict losses based on annualized losses or other mathematical functions.

SOURCE: Information provided by S. Cutter; http://sheldus.org

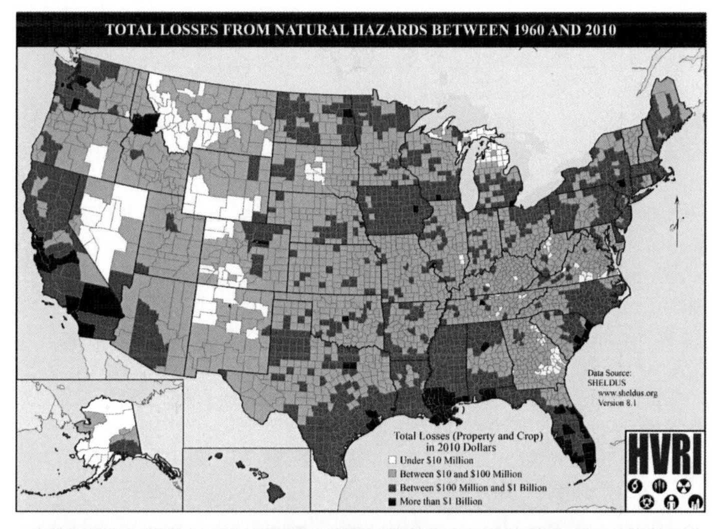

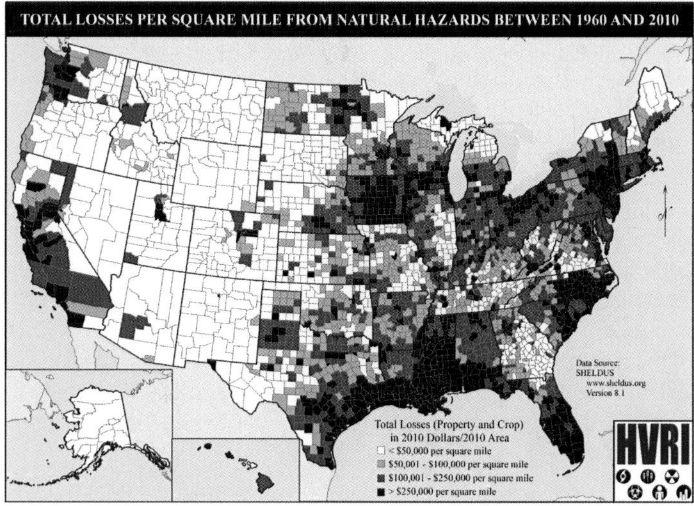

FIGURE 3.3 Geographic pattern of property and crop losses from natural hazards, 1960-2010. (a) Cumulative losses; (b) losses normalized by area (square mile). In addition to the clear concentration of losses along the Gulf Coast and southeastern coastal region, other concentrations of losses occur in California, western Washington, Mid-Atlantic (New Jersey, Pennsylvania, and New York), and in the upper Great Plains (North Dakota and Minnesota). The state-centric pattern in (a) for Iowa and Mississippi is partially a function of data-reporting biases in the source data. The overall pattern of losses in Idaho, and particularly in Lewis county, is not a function of a single extreme event, but instead a series of events which, over time, contributed to the total. Although significant, many of these individual events were not severe enough to warrant a presidential disaster declaration; yet over time, such repetitive losses affected the counties' abilities to respond and recover and led to millions of dollars in crop damages. HVRI = Hazards and Vulnerability Research Institute. Source: S. Cutter/HVRI.

Given that the past 50 years may not be a good indicator of future patterns in hazard losses, either for weather-related events likely to be impacted by climate change or for hazards with long return periods such as earthquakes, other probabilistic models can be used to predict the potential distribution or impact of future losses for the nation. FEMA's natural hazard loss estimation model, HAZUS, enables users to project losses for a community or region based on inputs about a specific event that is defined by the user. Alternatively, HAZUS can provide probabilistic loss estimates nationwide when the USGS probabilistic seismic source model is input. An example of the output of such modeling is the annualized earthquake losses by county for the United States (FEMA, 2008) (Figure 3.4).

Nationwide, the total modeled annualized loss of national building stock from earthquakes is about $5.3 billion.[2] If indirect business interruption were taken into account the economic losses from earthquakes would be even greater and more widely distributed. The map of total annualized earthquake losses shown in Figure 3.4a demonstrates that nearly the entire nation is subject to potential earthquake loss; however, the greatest risk exists along the West Coast. Los Angeles County alone accounts for 25 percent of the entire nation's annualized loss, which is not surprising considering the large number of active faults in the region and the fact that the population of this single county is greater than all but eight states in the country. California in total has about 66 percent of the nation's total annualized loss; the Pacific Northwest together with California encompasses about 77 percent of the nation's annualized earthquake loss.[3] The map of normalized AEL (ratio of total loss to millions of dollars of building inventory value) in Figure 3.4b highlights concentrated loss in regions of high seismic hazard outside of the West Coast: the Wasatch Front in Utah and extending north through the Rocky Mountains, as well as sites of historic earthquakes in the central and eastern United States for which there is geological evidence of repeated events over the past several thousand years (New Madrid, Missouri region; Charleston, South Carolina; and along the Saint Lawrence Seaway).

---

[2] http://www.fema.gov/plan/prevent/hazus/hz_aelstudy.shtm.
[3] http://www.fema.gov/plan/prevent/hazus/hz_aelstudy.shtm.

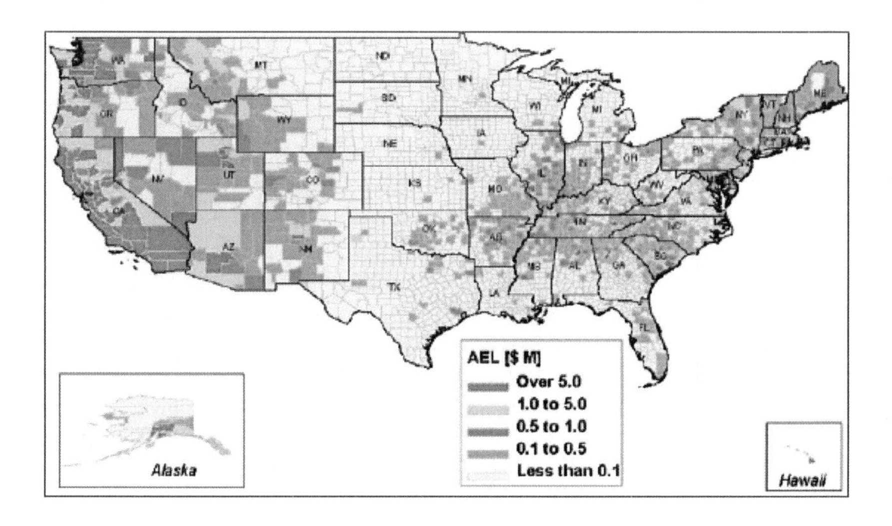

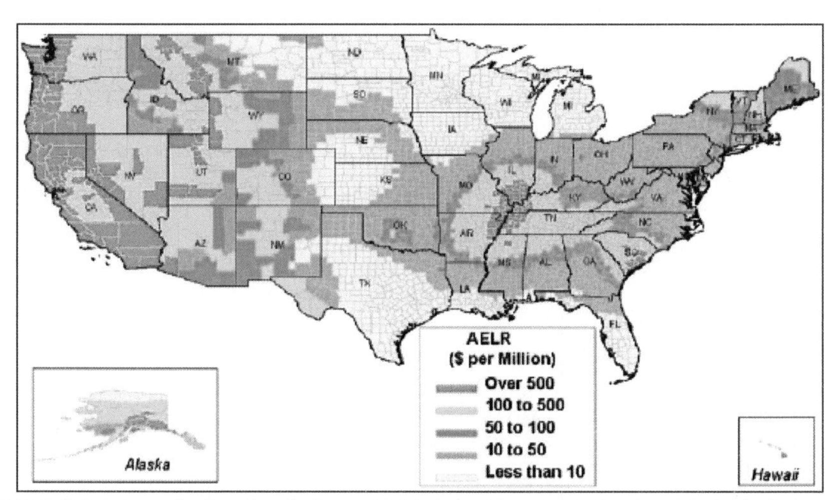

FIGURE 3.4 Annualized earthquake losses (AEL) derived from HAZUS using U.S. Geological Survey probabilistic seismic hazard assessment as input. (a) AEL (total dollar value loss of all structures included in the HAZUS exposure inventory); (b) normalized AEL (ratio of total loss to millions of dollars of building inventory value (the value of all buildings in the study area). Source: Federal Emergency Management Agency.

Human Losses and Loss-of-Life Data

Whereas national and global economic losses are growing annually, a positive development is that human losses (deaths, injuries, displacements) generally show the opposite tendency, especially in the developed world (Goklany, 2009). In the United States, the number of fatalities due to disasters has remained roughly steady between the 1990s and 2000s. In contrast, deaths and numbers of people affected by disasters continue to grow in the developing nations (IFRCRC, 2010); in fact, the number of people affected (those requiring immediate assistance, those who are injured, or those made homeless from the disaster) increased threefold during the first decade of the 21st century (IFRCRC, 2010).

The declining number of deaths from natural disasters in the United States and the rest of the developed world is mostly the result of improved building codes and construction practices, improved awareness about disaster risk, and more accurate forecasting and warning systems. Considerable research on disaster mortality has been conducted, especially on specific perils such as floods (Ashley and Ashley, 2008; Zahran et al., 2008), earthquakes (Shoaf et al., 1998), and severe weather (Ashley, 2007). Well-established research also exists on specific mortality-causing disasters such as Hurricane Andrew (Combs et al., 1996), the Northridge earthquake (Peek-Asa et al., 2000), the Chicago heat wave (Klinenberg, 2003), and Hurricane Katrina (Elder et al., 2007; Jonkman et al., 2008). Despite those significant efforts, however, research results on all-hazards mortality in terms of temporal and spatial patterns are few (Borden and Cutter, 2008; Thacker et al., 2008) and do not provide the quality and quantity of data necessary for understanding the overall human losses.

As was the case with economic losses, loss-of-life data for disasters can also be difficult to use and interpret (Box 3.5). NOAA and the Centers for Disease Control (CDC) are the primary natural-disaster-fatality sources in the federal government. Their data include direct and indirect fatalities related to a disaster event. Death certificates are the source of the input data in CDC's mortality databases. NOAA also records fatality statistics based on reports by local National Weather Service offices and the news media and then consolidates estimates into the monthly Storm Data. The CDC and NOAA fatality databases differ in the classification of perils, which deaths are counted, and the attribution of the death to a specific peril or place (Figure 3.5).

A further complication with mortality data is that most hazard-mortality research uses raw counts of fatalities that are not adjusted to either rates (deaths per population), densities (per unit area), or standardized mortality ratios (accounting for the age/sex structure and size of the population). This lack of refinement may present a very misleading indication of the nature of human losses from natural disasters, especially when attempting to examine regional variations. Moreover, the extreme variation in the scale of U.S. counties, both in terms of population and area, makes interpretation of county-level maps, such as those that illustrate this chapter, especially problematic. U.S. counties vary in population from less than 100 to roughly 10 million, and in area from less than 2 square miles to more than 150,000 square miles. However, counties remain the administrative unit for most hazard and risk management programs, and so we opt to report data at this level of resolution.

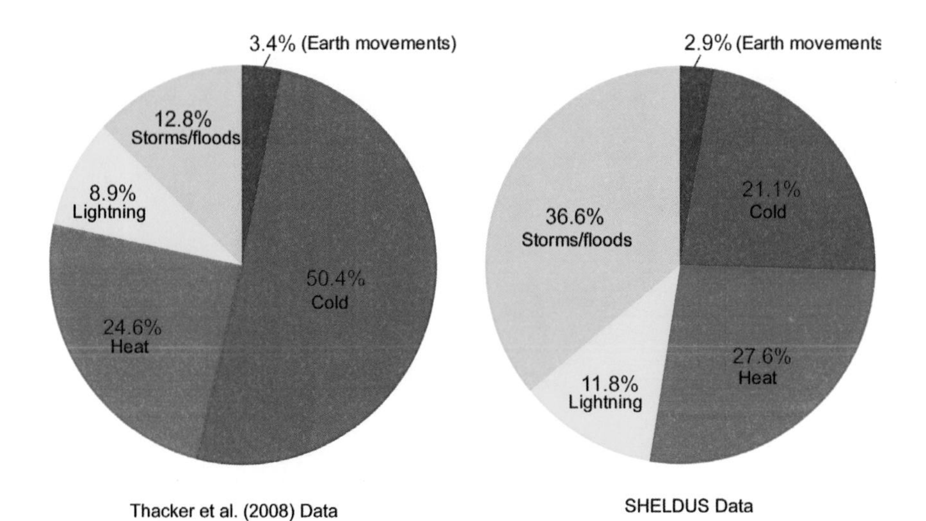

FIGURE 3.5 Hazard fatalities 1979-2004 compared by perils. Earth movements refers to earthquake-related fatalities and landslides. (a) Centers for Disease Control and Prevention (Thacker et al., 2008) and (b) National Oceanic and Atmospheric Administration (Borden and Cutter, 2008). CDC data are more likely to include urban and exposure deaths (heat and cold extremes), whereas NOAA data are biased toward more rural events such as lightning. The comparison illustrates the difficulty and inconsistency in data and recording the peril that contributed to the hazard fatality. Source: S. Cutter; complied from SHELDUS.

BOX 3.5
Problems with Collecting and Interpreting Disaster Fatality Data

Tracking deaths is relatively straightforward because all deaths are required by law to be reported. The difficulties with disaster fatality data are in how to attribute the cause of the death to a particular disaster or peril. This designation will vary depending on who is doing the reporting and recording on the death certificate. Attribution of the cause of death and the conditions contributing to it become highly subjective, and pronouncing physicians may have difficulties completely identifying the contributing conditions. For example, if a person has a heart attack while shoveling snow, the death may or may not be recorded as a disaster death depending on how the paperwork is completed. The cause of death would be a heart attack, but the contributors would be physical exertion due to the snowstorm. A further complication related to disaster fatality statistics is determining the location where the death occurred. A death certificate contains a place to fill in the geographic location of the initial injury (street, county, zip code, etc.). If left blank, the fatality is georeferenced to the place of residence of the deceased, or is recorded as the place where the death pronouncement was made (e.g., a hospital). For example, if a tourist from Arkansas died in a wildfire while on vacation in Colorado, the death could be recorded as a wildfire fatality in Arkansas (where the person lived), but it could also be listed for Colorado (where he or she died), depending on how the death certificate was completed. Finally, problems arise with the timing of the death. Many suicides and deaths related to toxic exposures post-Katrina were not recorded as related to Katrina. Deaths from toxic exposures experienced by first responders to the 9/11/2001 destruction of the World Trade Center towers in New York City are still occurring.

Patterns of Exposure and Population Growth

Population growth affects exposure to hazards for a variety of reasons. Understanding these population patterns through time allows assessment of some of the underlying socioeconomic or demographic changes that may contribute to the vulnerability of communities to disasters. Population growth or decline in a geographic location can also relate to infrastructure issues pertinent to the particular hazards associated with that region. For example, the new infrastructure needs (housing, roads, bridges) that growing communities need may require decisions to be made regarding land use and development in undeveloped areas. These undeveloped areas may include areas of natural defenses whose integrity may be important as a mitigation measure against existing natural hazards (see Chapter 2). The United States is experiencing a major transformation in population development patterns, which began in the 1970s with the movement of population out of the northern Rust Belt states into the south and southwest. This period saw a tremendous influx into coastal counties where approximately 53 percent of the U.S. population now resides and where about half of the nation's residential units are located (Crosset et al., 2004). Over the period from 2000 to 2010, the migration to the coastal counties has slowed somewhat (Figure 3.6) although growth in selected Florida counties exceeded 50 percent. Depopulation is another aspect that is visible in Figure 3.6, with much of the Great Plains showing decreasing populations. Also notable is the declining population in counties bordering the lower Mississippi River from southern Illinois to southern Mississippi (Mackun and Wilson, 2011). Large population losses during the decade occurred in Orleans and St. Bernard parishes in Louisiana (due to Hurricane Katrina), in Cameron Parish, Louisiana (due to Hurricane Ike), and in Issaquena and Sharkey counties in Mississippi due to poor economic conditions and a long history of population decline from the counties.

The aging of the U.S. population is also important to consider. The growing number of older adults who need more specialized care will require greater knowledge, expertise, equipment, and supplies during a disaster, particularly during an evacuation. This problem was very clearly evident in the hours and days that followed Hurricane Katrina, where responders were not prepared to handle the medical needs they encountered in older adults (NRC, 2011).

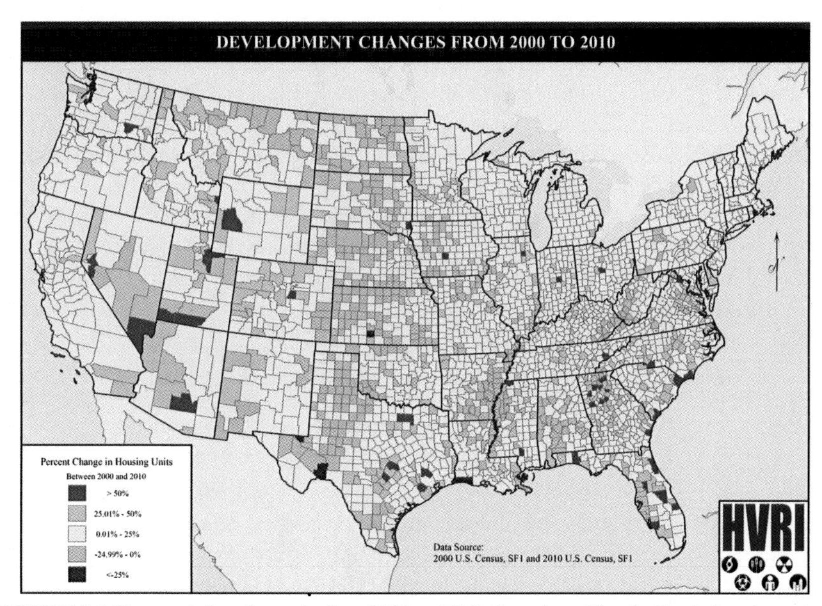

FIGURE 3.6 Changes in housing units from 2000 to 2010. Counties with a decline in housing units are shown in purple; areas with increasing housing units are shown in dark green. Source: S. Cutter/HVRI.

What is missing from this narrative is the overlay of the population shifts and residential housing units with consistent national probabilistic hazard maps (such as the USGS National Seismic Hazard Map and the FEMA flood zone maps) and with accurate mapping of both structural and social vulnerability. Although existing data allow discussion about increases and decreases in exposure, conclusions remain difficult to make regarding the effects on resilience of changes in populations in hazardous zones such as flood-prone areas or seismically active regions. Consistent multihazard data for the entire country calibrated from the local to the state level together with local- to regional-scale vulnerability data are needed to create true national risk maps to allow comparison of relative risks from different perils as different return periods. Such information could provide the basis for community prioritization of limited resources to expend on reducing risk and building resilience.

KNOWLEDGE AND DATA NEEDS

The lack of standardization of data on hazardous conditions, disaster losses, and impacts is a continuous challenge in the effort to understand and managing risk and increase disaster resilience. Hazard and disaster informatics is a relatively new scientific field, yet information derived from this area of study is critical to the national resilience efforts. The CDC is beginning to coordinate health and health-risk and emergency data from hospitals and health departments, but medical professionals are a long way from having the ability to access individual health records in an emergency, or recording losses of life and health accurately following a disaster. A number of NRC reports recognize the need addressing issues of hazard and disaster data collection, standardization, management, archiving, and sharing (NRC, 2006b, 2007a,b,c). Whether such principles are achieved through a nongovernmental panel looking at hazard and disaster informatics (NRC, 2006a) or through the formal establishment of a national loss inventory (Mileti, 1999; Cutter, 2001) or other mechanisms is open for discussion. What is not debatable is the criticality of the need to solve the disaster informatics issue.

SUMMARY AND RECOMMENDATION

The ability to measure and evaluate the assets of communities and to understand the economic and human value of resilience is critical to improving disaster resilience. The resources of a community involve more than the high-value essential assets such as hospitals and utilities. They also include other assets with high social, cultural, and environmental value, and so decision-making models developed by communities have to involve both quantitative and qualitative "valuation" of assets in order to prioritize resilience investments. Presently, little guidance exists for communities to understand how to place meaningful value on both their quantitative and qualitative assets.

In developing the case for enhancing resilience now and providing motivation for community decision makers to understand their inventory of assets and the ways in which they interact with one another, this chapter has also outlined the historical spatial and temporal patterns of economic and human disaster losses on communities in the United States. Although the data available to assess economic and human losses nationally are conservative and are neither comprehensive nor centrally archived for the nation, the historical patterns of economic losses from hazards and disasters in the United States appear to be increasing and will be expensive to absorb, if allowed to continue. A positive sign—a declining tendency in human losses (fatalities) from disasters in the United States and other developed countries—attests to the success of some resilience-building measures. Improved building codes, improved awareness, and more accurate forecasting and better warning systems are some of the factors researchers think may contribute to fewer fatalities from disasters.

However, changing patterns of hazards as well as changes in population and vulnerability affect economic and human loss patterns. Attempts to improve resilience of individual communities and the nation require more consistent hazard and risk assessments supported by consistent and centrally available disaster loss data. Accurate loss and casualty data on past disasters enable researchers to better constrain the factors controlling the structural and social vulnerability of communities and also enable practitioners to quantitatively calibrate risk/loss models and make more accurate predictions of future losses. This lack of data compromises the ability of communities to make informed decisions about resilience-building strategies. Importantly, the need for resilience-building strategies continues even if historical patterns of loss begin to improve.

Recommendation: **A national resource of disaster-related data should be established that documents injuries, loss of life, property loss, and impacts on economic activity. Such a database will support efforts to develop more quantitative risk models and better understand structural and social vulnerability to disasters.** To improve access to these data, the principle of open access should be recognized in all relevant federal data management policies. The data should be made accessible through an Internet portal maintained either by a designated agency or by an independent entity such as a university. The National Science and Technology Council (NSTC) would be an appropriate entity to convene federal and state agencies, private actors, nongovernmental organizations, and the research community to develop strategies and policies in support of these data collection and maintenance goals.

Such a data inventory would reconcile and integrate the fragmented federal datasets on disasters and losses; serve as a national data archive for historic hazard events and loss data; assist in the development of better loss metrics; and provide the evidentiary basis for potentially evaluating resilience interventions. Federal agencies, private actors, and the research community working in concert would improve post-event data collection and public access to such data, would determine essential data, standards, and protocols to employ, and determine which agencies are best positioned to collect and archive specific data on the impacts of hazards. Such an approach helps to avoid duplication of efforts. Likely federal actors include FEMA, NOAA, CDC, USGS, the U.S. Forest Service, and the U.S. Army Corps of Engineers. Biennial status reports coordinated by the NSTC on the nation's resilience could be based in part on an analysis of these data, and could include priorities for future data collection and dissemination. At the same time, data on resilience are also lacking. Chapter 4 discusses specific ways in which resilience can be measured and used as a basis for such status reports.

REFERENCES

Ashley, S. T., and W. S. Ashley. 2008. Flood fatalities in the United States. *Journal of Applied Meteorology and Climatology* 47(3):805-818.

Ashley, W. S. 2007. Spatial and temporal analysis of tornado fatalities in the United States: 1880-2005. *Weather and Forecasting* 22(6):1214-1228.

Borden, K., and S. L. Cutter. 2008. Spatial patterns of natural hazard mortality in the United States. *International Journal of Health Geographics* 7:64. Available at http://www.ijhealthgeographics.com/content/7/1/64.

Brooks, H. E., and C. A. Doswell III. 2001. Normalized damage from major tornadoes in the United States: 1890-1999. *Weather and Forecasting* 16(1):169-176.

Combs, D. L., R. G. Parrish, S. J. N. McNabb, and J. H. Davis. 1996. Deaths related to Hurricane Andrew in Florida and Louisiana, 1992. *International Journal of Epidemiology* 25(3):537-544.

Crosset, K. M., T. J. Culliton, P. C. Wiley, and T. R. Goodspeed. 2004. Population Trends Along the Coastal United States: 1980-2008. NOAA National Ocean Service. Available at http://oceanservice.noaa.gov/programs/mb/pdfs/coastal_pop_trends_complete.pdf.

Cutter, S. L., ed. 2001. *American Hazardscapes: The reRgionalization of Hazards and Disasters.* Washington, DC: Joseph Henry Press.

Elder, K., S. Xirasagar, N. Miller, S. A. Bowen, S. Glover, and C. Piper. 2007. African Americans' decision not to evacuate New Orleans before Hurricane Katrina: A qualitative study. *American Journal of Public Health* 97(1):S124-S129.

FEMA (Federal Emergency Management Agency). 2008. HAZUS® MH Estimated Annualized Earthquake Losses for the United States. FEMA 366. Available at http://www.fema.gov/library/viewRecord.do?id=3265.

Gall, M., K. Borden, and S. L. Cutter. 2009. When do losses count? Six fallacies of natural hazards loss data. *Bulletin of the American Meteorological Society* 90(6):799-809.

Gall, M., K. A. Borden, C. T. Emrich, and S. L. Cutter. 2012. The unstable trend of natural hazard losses in the United States. *Sustainability* 3:2157-2181.

Giesecke, J., W. Burns, A. Barrett, E. Bayrak, A. Rose, P. Slovic, and M. Suher. 2011. Assessment of the regional economic impacts of catastrophic events: A CGE analysis of resource loss and behavioral effects of a radiological dispersion device attack scenario. *Risk Analysis* 33(4):583-600.

Goklany, I. M. 2009. Deaths and death rates from extreme weather events: 1900-2008. *Journal of American Physicians and Surgeons* 14(4):102-109.

Heinz Center. 1999. *The Hidden Costs of Coastal Hazards: Implications for Risk Assessment and Mitigation.* Washington, DC: Island Press.

IFRCRC (International Federation of Red Cross and Red Crescent Societies). 2010. *World Disasters Report 2010: Focus on Urban Risk.* Available at http://www.ifrc.org/Global/Publications/disasters/WDR/wdr2010/WDR2010-full.pdf.

Jonkman, S. N., B. Maaskant, E. Boyd, and M. Levitan. 2008. Loss of life caused by the flooding of New Orleans after Hurricane Katrina: A preliminary analysis of the relationship between flood characteristics and mortality. In *4th International Symposium on Flood Defence.* Toronto, Ontario: Institute for Catastrophic Loss Reduction, pp. 96-1 to 99-9. Available at http://www.hkv.nl/documenten/Loss_of_life_caused_by_the_flooding_of_New_Orleans_after_hurricane_Katrina_BM(1).pdf.

Klinenberg, E. 2003. *Heat Wave: A Social Autopsy of Disaster in Chicago.* Chicago: University of Chicago Press.

Mackun, P., and S. Wilson. 2011. Population Distribution and Change: 2000 to 2010. Census Brief C2010BR-01. U.S. Bureau of the Census. Available at http://www.census.gov/prod/cen2010/briefs/c2010br-01.pdf.

Miller, S., R. Muir-Wood, and A. Boissonnade. 2008. An exploration of trends in normalized weather-related catastrophe loss. In *Climate Extremes and Society.* Cambridge, UK: Cambridge University Press, pp. 225-247.

Mileti, D. 1999. *Disasters by Design: A Reassessment of Natural Hazards in the United States.* Washington, DC: Joseph Henry Press.

Multihazard Mitigation Council. 2005. *Natural Hazard Mitigation Saves: An Independent Study to Assess the Future Savings from Mitigation Activities, Vol. 1, Findings, Conclusions, and Recommendations.* Washington, DC: National Institute of Building Sciences. Available at www.nibs.org/MMC/MitigationSavingsReport/Part1_final.pdf.

Munich RE. 2012. *Topics Geo 2011*. Munich: Münchener Rückversicherungs-Gesellschaft.

NRC (National Research Council). 1999. *The Impacts of Natural Disasters: A Framework for Loss Estimation.* Washington, DC: National Academy Press.

NRC. 2006a. *Facing Hazards and Disasters: Understanding Human Dimensions.* Washington, DC: The National Academies Press.

NRC. 2006b. *Improved Seismic Monitoring—Improved Decision-Making: Assessing the Value of Reduced Uncertainty*, Washington, DC: The National Academies Press.

NRC. 2007a. *National Land Parcel Data: A Vision for the Future.* Washington, DC: The National Academies Press.

NRC. 2007b. *Successful Response Starts with a Map: Improving Geospatial Support for Disaster Management.* Washington, DC: The National Academies Press.

NRC. 2007c. *Tools and Methods for Estimating Populations at Risk from Natural Disasters and Complex Humanitarian Crises.* Washington, DC: The National Academies Press.

NRC. 2011. *Increasing National Resilience to Hazards and Disasters: The Perspective from the Gulf Coast of Louisiana and Mississippi.* Washington, DC: The National Academies Press.

Peek-Asa, C., J. F. Kraus, L. B. Bourque, D. Vimalachandra, J. Yu, and J. Abrams. 1998. Fatal and hospitalized injuries resulting from the 1994 Northridge earthquake. *International Journal of Epidemiology* 27(3):459-465.

Pielke, R. A. J., and C. W. Landsea. 1998. Normalized hurricane damages in the United States: 1925-95. *Weather and Forecasting* 13:621-631.

Rose, A. 2004. Economic principles, issues, and research priorities in natural hazard loss estimation. In *Modeling the Spatial Economic Impacts of Natural Hazards,* Y. Okuyama and S. Chang, eds. Heidelberg: Springer, pp. 13-36.

Rose, A., and S. B. Blomberg, 2010. Total economic impacts of a terrorist attack: Insights from 9/11. *Peace Economics, Peace Science, and Public Policy* 16(1), Article 2.

Rose, A., K. Porter, N. Dash, J. Bouabid, C. Huyck, J. C. Whitehead, D. Shaw, R. T. Eguchi, C. Taylor, T. R. McLane, L. T. Tobin, P. T. Ganderton, D. Godschalk, K. Tierney, and C. T. West. 2007. Benefit-cost analysis of FEMA hazard mitigation grants. *Natural Hazards Review* 8(4):97-111.

Shoaf, K. I., L. H. Nguyen, H. R. Sareen, and L. B. Bourque, 1998. Injuries as a result of California earthquakes in the past decade. *Disasters* 32(2):303-315.

Sutter, D., D. DeSilva, and J. Kruse. 2009. An economic analysis of wind resistant construction. *Journal of Wind Engineering and Industrial Aerodynamics* 97:113-119.

Thacker, M. T. F., R. Lee, R. I. Sabogal, and A. Henderson. 2008. Overview of deaths associated with natural events, United States, 1979-2004. *Disasters* 32(2):303-315.

Zahran, S., S. D. Brody, W. G. Peacock, A. Vedlitz, and H. Grover. 2008. Social vulnerability and the natural and built environment: A model of flood casualties in Texas. *Disasters* 32(3):537-560.

"(We) look at trends in the New Orleans area across 3 decades to get the entire view of the health and vitality of the city as a measure of the city's resilience..."
Allison Plyer, Greater New Orleans Community Data Center, January 20, 2011

4

Measuring Progress Toward Resilience

THE NEED FOR METRICS AND INDICATORS

The committee recognized early on in its discussions that the study's focus on improving resilience necessitates measurement, a position also indicated in the study's Statement of Task (see Chapter 1). Measurement is essential for several reasons. First, it would be impossible to identify the priority needs for improvement without some numerical means of assessment. Second, a system of measurement is essential if progress is to be monitored. Third, any effort to compare the benefits of increasing resilience with the associated costs requires a basis of measurement. Establishing a baseline or reference point from which changes in resilience can be measured, combined with a regular system of monitoring to track changes through time, is also necessary. However, the measurement of a hard-to-define concept is necessarily difficult, requiring not only an agreed-upon metric, but also the data and algorithms needed to compute it. Resilience also includes human (social) and physical (infrastructure, natural environment) components that add complexity and challenges in finding metrics that cover this range of factors. This chapter discusses some of the more important principles and issues connected with measuring resilience. It examines the available methods, data, and tools, and makes recommendations designed to implement one type of measuring system for resilience.

One national-scale metric of resilience could be the dollar amount (per capita) of federal assistance spent annually for disasters, with the measure for resilience being whether this dollar amount flattens or declines (potentially indicating increasing resilience) or continues its steady growth (potentially indicating that resilience is not increasing, or is not increasing at a significant rate nationally). While imperfect, such an indicator provides a valuable synoptic, national picture, but other metrics would be required to measure the progress of individual communities.

Metrics are an important tool of administration. They allow targets to be established and set clear goals for improvement. The very act of defining a metric, and the discussions that ensue about its structure, help a community to clarify and formalize what it means by an abstract concept, thereby raising the quality of debate. The general concept of resilience is one with which most people are familiar, but resilience is not something that communities have much experience in measuring. Resilience is also clearly influenced by multiple factors, making precise measurement very difficult. This immediately suggests a strategy of combining various factors, using appropriate weights, into a composite index. The set of factors, how they are measured, the weights given to each factor, and the operations used to combine them into a composite index all present issues that can be the subject of lengthy debates and contention. At the same time, the translation of an abstract concept into a rigorous procedure for measurement—the formalization of the concept—allows for monitoring, the comparison of progress in different communities, and the prioritization of actions and investments, all of which can be extremely helpful. The effects of actions and policy changes can then be monitored through time to produce more desirable outcomes in the future by comparing improvements in resilience that result from those actions to what was promised or predicted, iteratively modifying actions and policies, and perhaps recalibrating metrics.

To be useful in this context, a resilience metric needs to be open and transparent, so that all members of a community understand how it was constructed and computed. It needs to be replicable, providing sufficient detail of the method of determination of a community's resilience so that it can be checked by anyone using the same data. It must also be well documented and simple enough to be used by a wide range of stakeholders.

Metrics may be quantitative, but metrics with no more than *ordinal* properties still allow resilience to be ranked and progress to be monitored. For example, a metric might set the qualitative levels "unsatisfactory," "marginal," and "satisfactory" resilience, without specifying quantitative measures or ranges for each level, as long as the procedure for arriving at a rating was open, transparent, and replicable. A scale similar to those used in academic report cards with designations of A, B, C, D, or F could also be used to indicate progress. In recent years, much of this process of defining a metric has been the subject of extensive research, often under the rubric of multicriteria decision making (MCDM). Many of these methods have been devised for problems embedded in geographic space, such as the selection of a site for a new public facility, or of a route for a new highway. The geospatial nature of such problems raises additional issues such as estimating environmental, social, and economic impacts of site selection for the new development and the way in which the necessary data to gauge these impacts can be incorporated into a collective planning process, as several texts make clear (see, e.g., Massam, 1993;

Malczewski, 2010). The methods deal effectively with the disparate views of stakeholders, allowing consensus to emerge and measuring the degree to which consensus exists. For example, the analytical hierarchy process (Saaty, 1988) is a much-applied method for reconciling divergent views in the creation of a consensus metric.

Many of these principles are illustrated by the well-known LEED (Leadership in Energy and Environmental Design; Box 4.1) process, released by the U.S. Green Building Council in March 2000. By providing an open forum for the measurement of environmental sustainability of buildings, LEED has provided an important tool for promoting and achieving energy efficiency. LEED was a bottom-up initiative without any initial endorsement from government agencies. It has gained popularity in engineering and architectural design as an added value to building occupants and to the environment in general. It has also become a trademark of socially conscious organizations in the private sector. The committee was struck by the impact LEED has had and seeks to emulate its success by envisioning a similar strategy for the measurement of resilience, laid out in the final section of this chapter.

BOX 4.1
Leadership in Energy and Environmental Design

LEED, or Leadership in Energy and Environmental Design, is an internationally recognized green-building certification system. Developed by the U.S. Green Building Council (USGBC) in March 2000, LEED is a framework for building owners and operators that allows identification and implementation of green building design, construction, operations, and maintenance.

LEED promotes sustainable building and development practices through a set of rating systems that recognize building projects that have adopted strategies for better environmental and health performance. The LEED rating systems are developed through an open, consensus-based process led by LEED committees comprising groups of volunteers from across the building and construction industry. Key elements of the process of developing LEED rating systems include a balanced, transparent committee structure, technical advisory groups for scientific consistency and rigor, opportunities for stakeholder comment, member ballot of new rating systems, and fair and open appeals.

LEED can apply to all building types, whether commercial or residential. LEED works throughout the building life cycle from design and construction through to tenant fitout and retrofit. LEED for Neighborhood Development is designed to allow the benefits of LEED to extend beyond a single building and into the neighborhood it serves.

SOURCE: http://www.usgbc.org/DisplayPage.aspx?CMSPageID=1988.

While LEED focuses primarily on buildings, the thrust of this chapter's discussion is on the resilience of communities and their complexities. For example, a metric of the overall resilience of an entire city may mask substantial variations within the city. Carried to an extreme, we might conceive of resilience as varying continuously over the Earth's surface, similar to the way elevation varies, and scale-dependent in both space and time. Moreover, resilience is a function of many factors, not all of which may be the same for all people, even when those people occupy the same location.

Problems such as these are familiar to geographers and others who work with geospatial data, and are commonly termed the *Modifiable Areal Unit Problem* (see, e.g., Longley et al., 2011). Such problems arise when the results of an analysis, such as the measurement of resilience, depend on the areas used for the analysis. We might find, for example, that neighborhoods in some areas of New Orleans are substantially more resilient than other neighborhoods and that New Orleans as a whole has a resilience in the middle of the range, when compared with other places. By selectively lumping neighborhoods together, in other words, by modifying the areal units in a process similar to gerrymandering electoral districts, one could produce a map that sharply and misleadingly contrasts highly resilient areas and much less resilient areas.

The committee recognized the need to address this problem in any recommended system of measurement. The key is the concept of community, and its requirements of self-identification and mutual affinity, allowing a community, its members, and its boundary to be treated as an existing, well-defined area. In this sense a neighborhood, a town, or an entire city might all qualify as communities; and a community need not be formally recognized as an administrative unit, or precisely defined by a boundary on the Earth's surface. Any individual might belong to more than one community, each with its own measurement of resilience; a New Orleans resident might live in a highly resilient neighborhood, but in a city of relatively low resilience. With this principle as its foundation, and no possibility of arbitrary or selective gerrymandering, the process of measurement of community resilience becomes much more straightforward. Essentially, and recalling a long-recognized duality in geography and related disciplines (e.g., Tuan, 2007), resilience needs to be addressed by reference to *place* and not *space*.

MEASURES OF U.S. NATIONAL RESILIENCE

Many organizations have tackled the problem of measuring resilience, or its close relative vulnerability, for the United States. This section reviews many of these efforts, choosing specific representative examples for detailed discussion.

Coastal Resilience Index

The Coastal Resilience Index, cosponsored by the Louisiana Sea Grant, Mississippi-Alabama Sea Grant Consortium, and the National Oceanic and Atmospheric Administration Gulf Coast Services Center (Emmer et al., 2008), provides an example of a community-based approach to developing an index of resilience to storm events through self-assessment. It adapts the principles outlined by FEMA (2001) to the specific needs of coastal hazards and operationalizes them into an ordinal metric.

The community is first asked to identify two scenarios from memory: a "bad storm" and a "worst storm." Critical infrastructure and facilities are then evaluated: Were they impacted in either or both of the scenarios, and were they functioning afterward? Critical infrastructure includes the wastewater treatment system, the power grid, the water purification system, and transportation/evacuation routes. Critical facilities include government buildings, law enforcement buildings, fire stations, communication offices, the emergency operations center, evacuation shelters, hospitals, and critical record storage. The community is encouraged to expand these lists as appropriate. The numbers of critical infrastructure elements and critical facilities that continued to function after the scenarios are then totaled.

In the next step, the community is asked to assess whether various elements of its transportation system will be restored within 1 week after a "bad storm," and again to total the number of such elements. The third step asks for information on the participation of the community in various plans and agreements, and whether it has key personnel in place with responsibility for disaster-related matters. The number of positive responses is counted. Step 4 yields a total for ongoing mitigation measures, Step 5 addresses business plans for the recovery of retail stores, and Step 6 asks about social networks and civic organizations.

The totals in each step are next transformed to Low, Medium, and High categories based on specified ranges—for example, to gain a High rating on critical infrastructure the community must have agreed that 100 percent of its elements would be functioning after a disaster. No weights are applied to each element; rather, the community is asked simply to count. The result is a total of seven metrics (two from Step 1 and one from each of the subsequent steps). The community is advised to treat these as separate indicators and not to attempt to combine them into a single metric.

The Low, Medium, and High resilience ratings are then converted into an overall state-of-the-community resilience for a specific category, along with some estimate of the time it would take for reoccupation of the community after the disaster: more than 18 months for a Low rating; less than 2 months for a Medium rating; and minimal impact for a High rating.

Argonne National Laboratory Resilience Index

A very different approach to measuring the resilience of critical infrastructure is described by Fisher et al. (2010), the result of a project conducted by Argonne National Laboratory in collaboration with the U.S. Department of Homeland Security's Protective Security Coordination Division. Data are gathered at critical infrastructure facilities by trained interviewers known as Protective Security Advisors (PSAs). The interviews use an Infrastructure Survey Tool covering roughly 1,500 variables that cover six major physical and human components (physical security, security management, security force, information sharing, protective measures assessment, and dependencies) that are themselves broken down into 42 components. The approach is used for one or several types of critical infrastructure or key resource sector (banking and finance, dams, energy, etc.). Data are subjected to an elaborate, six-step process of quality control involving review by experts in critical infrastructure protection.

A five-stage aggregation process is then used to combine the items into a single Resilience Index (called the Protective Measure Index PMI) that ranges from 0 (lowest resilience) to 100 (highest resilience) for a given critical infrastructure or key resource sector and for a given threat. Each of the stages takes a subset of items at that stage and combines them using weights to obtain a single index for the next stage. From roughly 1,500 items at Level 5, this process results in 47 composite scores at Level 2, three at Level 1, and finally a single score. At Level 2, 18 of the 47 measures contribute to Robustness at Level 1, five to Recovery at Level 1, and 24 to Resourcefulness at Level 1. At each stage, every contributing measure is multiplied by a weight, and the products are summed to obtain the PMI composite index. Weights are obtained by analyzing the opinions of experts, using the MCDM methods of Keeney and Raiffa (1976). PMI ratings by sector (e.g., commercial facilities, energy, transportation, water) may help in identifying the infrastructure facility that is weakest in relation to one or several threats.

In contrast to the bottom-up elements of the Coastal Resilience Index, this approach is almost entirely top down, reflecting the need of a national program to be uniform and universal in its approach. There is no possibility of adaptation to local needs, by modifying either the set of data items or the weights, both of which are prescribed. The index is entirely concerned with critical infrastructure, such a narrow focus being more conducive to a rigorous, quantitative approach. Nevertheless, justifying universal weights resolved to three decimal places is difficult given the inherent vagueness of the concept of resilience and its essential components, and uncertainties over the exact nature of threats.

Social Vulnerability Index (SoVI®)

Social vulnerability is the susceptibility of a population to harm from a natural hazard and examines those characteristics of the population that influence their resilience. Vulnerable populations may be less resilient to hazards and disasters than other parts of the population, may need special assistance in preparing for, responding to, and recovering from disasters, and may be more susceptible to economic or other impacts from an event. Social vulnerability is place-based and context-specific, and helps explain why some portions of the country or community experience a hazard differently, despite having the same exposure. Income is but one variable that is often associated with vulnerable populations, and income levels clearly vary by race and ethnicity (Figure 4.1). Other vulnerable populations may include special-needs populations such as residents with physical or mental impairments, the elderly, the young, and those with limited access to transportation (see also Chapter 5).

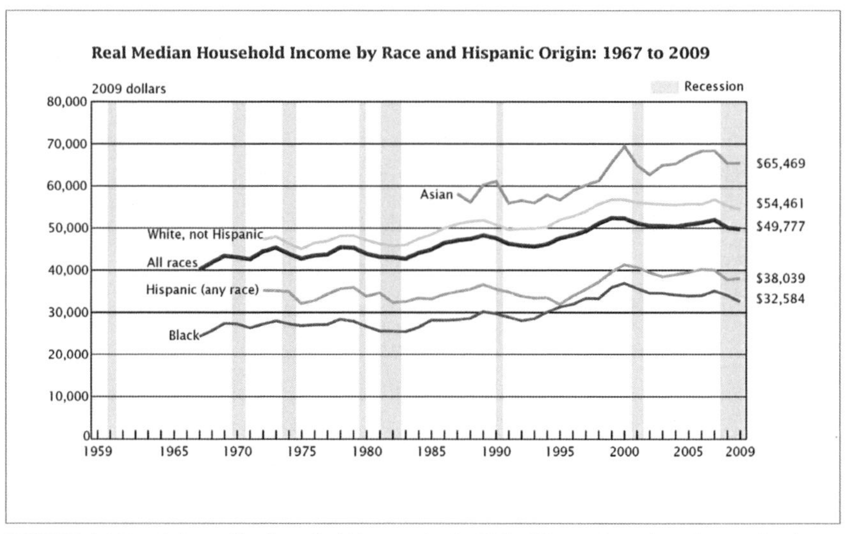

FIGURE 4.1 Trends in median household income in the United States. Data show income level variations by race and ethnicity. Source: U.S. Census Bureau.

Social vulnerability helps us to understand the inequalities in disaster impacts and is a multiattribute concept that includes socioeconomic status, race and ethnicity, gender, age, housing tenure, and so forth and how these factors influence a community's resilience (Mileti, 1999; Heinz Center, 2002; NRC, 2006). Social vulnerability can change over time and across space (Cutter and Finch, 2008) and can be measured both qualitatively and quantitatively (Birkmann, 2006; Phillips et al., 2010).

Social vulnerability metrics are increasing in sophistication and usage in both research and practice. Among the best known is the Social Vulnerability Index (SoVI®), a metric that permits comparisons of places (block groups, census tracts, metropolitan areas, counties) (Cutter et al., 2003; Box 4.2). Mapping SoVI® scores illustrates the extremes of social vulnerability—those places with very high values (the most vulnerable), and those with relatively low values (the least vulnerable) (Figure 4.2). SoVI® captures the multidimensional nature of social vulnerability—vulnerability that exists prior to any hazard or disaster event. In addition to describing the relative level of social vulnerability, the metric also enables the examination of those underlying dimensions that are contributing to the overall score such as age disparities, socioeconomic status, employment, and special-needs populations.

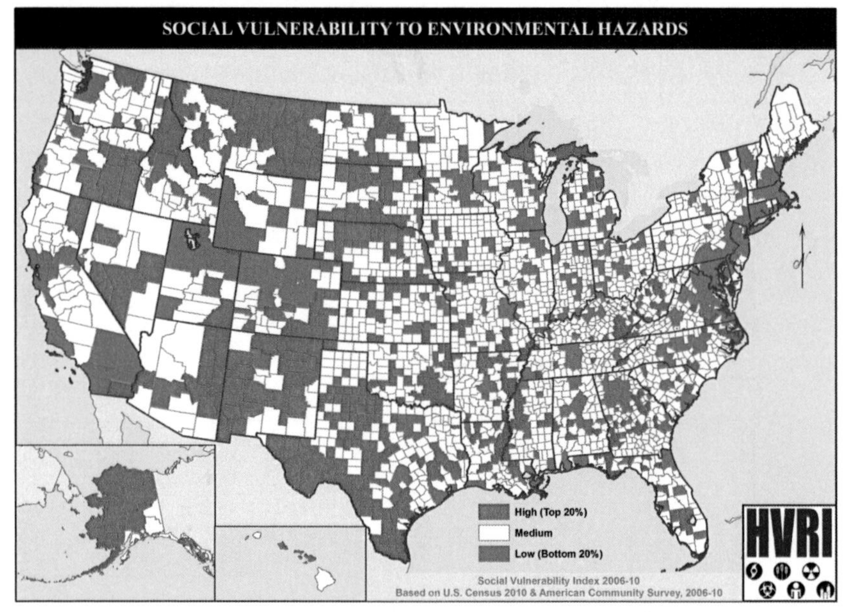

FIGURE 4.2 Social Vulnerability Index, 2006-2010. Areas in red denote higher levels of social vulnerability relative to other counties, whereas counties in blue show lower levels of social vulnerability. Mapping by standard deviations (represented here as top and bottom 20 percent) shows the extremes of the distribution, which is of greatest interest. HVRI = Hazard and Vulnerability Research Institute.
Source: S. Cutter/HVRI.

BOX 4.2
The Social Vulnerability Index (SoVI®)

SoVI® is a statistically derived comparative metric to illustrate the variability in capacity for preparedness, response, and recovery at county and subcounty levels of geography. Using census data, SoVI® synthesizes 32 different variables, using a principal components analysis and expert judgment, into a single composite value, which is then mapped to illustrate differences between places. Several factors consistently appear in the results of these analyses, including socioeconomic status, elderly, and gender; however, the relative importance of these factors is observed to be place specific. Since its inception, SoVI® has been used by emergency planners as part of their state hazard mitigation planning (South Carolina, California, and Colorado) and has been incorporated into a number of digital products including the National Oceanic and Atmospheric Administration's Coastal Services Digital Coast.
(http://www.csc.noaa.gov/digitalcoast/tools/slrviewer/index.html). See http://sovius.org for more details and applications.

Baseline Resilience Indicator for Communities

A new composite indicator called the Baseline Resilience Indicator for Communities (BRIC) was introduced to measure community resiliency (Cutter et al., 2010). BRIC acknowledges that resilience is a multifaceted concept with social, economic, institutional, infrastructural, ecological, and community components. The composite indicator is calculated as the arithmetic mean of five subindexes related to social, economic, institutional, infrastructural, and community resilience; ecological resilience is not included in the 2010 formulation. Each subindex is normalized so that the final indicator varies between 0 and 1.

Cutter et al. (2010) proposed several applications of the proposed method to communities at different scales. An interesting case study relates to the spatial distribution of disaster resilience over 736 counties within FEMA Region IV (Figure 4.3). A second example deals with determining the resilience score of three metropolitan areas: Gulfport-Biloxi, Charleston, and Memphis. Both case studies show a clear ability to identify least-resilient areas at different geographic scales using an empirically based descriptive approach.

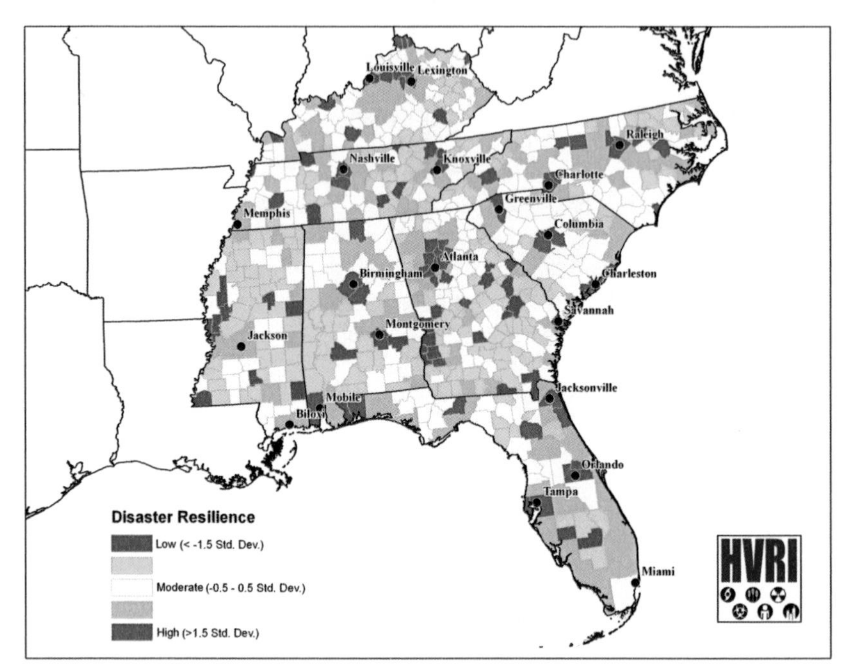

FIGURE 4.3 Spatial resolution of disaster resilience for FEMA Region IV. Source: S. Cutter/HRVI.

SPUR Model

The San Francisco Planning and Urban Research Association (SPUR) developed a set of metrics for measuring the resilience of the Bay Area with respect to earthquakes (SPUR, 2008). The process begins with the definition of an "expected earthquake," defined as one "that can reasonably be expected to occur once during the useful life of a structure or system," and in operation is one with a 10 percent probability of occurrence in a 50-year period. In the SPUR methodology, specific recovery objectives are defined in distinct time frames (Table 4.1): hours (3 to 72), days (30 to 60), and months (4 to 36). These target states of recovery and their time frames include those for hospitals, police and fire, the emergency operations center, transportation systems and utilities, airports, and neighborhood retail businesses, offices, and workplaces. Five categories of performance are defined for buildings ranging from A (safe and operational) to E (unsafe). Significantly, the goal for San Francisco was to have 95 percent of residents sheltering in place with 24 hours, requiring Category B performance for buildings. Although not all utilities might be functioning within 24 hours, the goal was to keep citizens in their homes and in their neighborhoods. The table provides the target states of recovery for San

Francisco's buildings and infrastructure together with an assessment of the current status for each of 31 distinct criteria. The gap between desired performance and current status highlights which areas need most work. No attempt is made in the model to collapse the criteria into a single metric. This approach provides a useful template that could be applied to an entire city, or to any neighborhood or community for use in defining their critical criteria for recovery, creating a timeline using performance objectives to achieve it, all in support of longer-term resilience goals.

TABLE 4.1 SPUR Model of Measuring Recovery from Earthquakes

TARGET STATES OF RECOVERY FOR SAN FRANCISCO'S BUILDING AND INFRASTRUCTURE

INFRASTRUCTURE CLUSTER FACILITIES	Event Occurs	Phase 1 Hours			Phase 2 Days		Phase 3 Months		
		4	24	72	30	60	4	36	36+
CRITICAL RESPONSE FACILITIES AND SUPPORT SYSTEMS									
Hospitals								X	
Police and fire stations			X						
Emergency operations center	X								
Related utilities						X			
Roads and ports for emergency				X					
CalTrain for emergency traffic					X				
Airport for emergency traffic				X					
EMERGENCY HOUSING AND SUPPORT SYSTEMS									
95% residence shelter-in-place								X	
Emergency Responder Housing				X					
Public shelters							X		
90% Related Utilities								X	
90% roads, port facilities, and public transit							X		
90% Muni and BART Capacity						X			
HOUSING AND NEIGHBORHOOD INFRASTRUCTURE									
Essential city service facilities							X		
Schools							X		
Medical provider offices								X	
90% neighborhood retail services									X
95% of all utilities								X	
90% roads and highways						X			
90% transit						X			
90% railroads							X		
Airport for commercial traffic					X				
95% transit							X		
COMMUNITY RECOVERY									
All residences repaired, replaced or relocated									X
95% neighborhood retail businesses open								X	
50% offices and workplaces open									X
Non-emergency city service facilities									
All businesses open									X
100% utilities									X
100% highway and roads									X
100% transit									X

Source: *SPUR Urbanist*, February 2009

The "x's" in the chart to the right indicate SPUR's best educated guesses about current standards for recovery times. The shaded areas represent the goals — targets based on clearly stated performance measures (see next page) — for recovery times for the city's buildings and lifelines. The gaps between "x's" and shaded boxes represent how far we are from meeting resiliency targets.

TARGET STATES OF RECOVERY

Performance Measure	Description of usability after expected event: BUILDINGS	LIFELINES
(lightest shading)	Category A: Safe and operational	
(light shading)	Category B: Safe and usable during repairs	100% restored in 4 hours
(medium shading)	Category C: Safe and usable after moderate repairs	100% restored in 4 months
(dark shading)	Category D: Safe and usable after major repairs	100% restored in 3 years
X	Expected current status	

Note: The table provides a useful template for identifying critical areas for recovery, which could provide the basis for establishing resilience goals. Source: C. Poland/SPUR.

Other Models and Metrics

Many other models and metrics have been developed for measuring progress toward resilience. A number of these are listed and described briefly Table 4.2. This table and Table 4.3 provided examples that the committee used to develop the perspectives presented in Section 4.4.

Table 4.2 Additional Models and Metrics of U.S. National Resilience

Community Assessment of Resilience Tool (CART)	A product of the National Consortium for the Study of Terrorism and Responses to Terrorism (START), CART is "a community intervention that includes a survey instrument, focus groups script, and process for assessing and building community resilience to disasters. Seven community capacity and competence attributes have been identified, refined, revised, and re-organized into four interrelated domains thought to affect community resilience to disasters: Connection and Caring, Resources, Transformative Potential, and Disaster Management. The current CART survey instrument consists of 21 core community resilience items along with demographics of respondents and additional questions dealing with issues of particular interest to participating organizations. The survey can be administered in person, over the telephone, by mail, or online. Results are used to develop a community profile from the perspective of respondents, a community intervention designed to measure and enhance community resilience" (START, 2011).
Community Resilience System (CRS)	CRS has been developed by the Community and Regional Resilience Initiative (CARRI, 2011). It includes six stages: Engage Community Leadership at Large, Perform Resilience Assessment, Develop Shared Community Vision, Action Planning, Establish Mechanism to Implement Plan and Sustain Program, and Evaluate and Review the Community's Resilience Program.
T*H*R*I*V*E	The Toolkit for Health and Resilience in Vulnerable Environments (T*H*R*I*V*E) was developed by Prevention Institute under contract to the U.S. Office of Minority Health. It provides a toolkit "to help communities bolster factors that will improve health

	outcomes and reduce disparities experienced by racial and ethnic minorities. T*H*R*I*V*E provides a framework for community members, coalitions, public health practitioners, and local decision makers to identify factors associated with poor health outcomes in communities of color, engage relevant stakeholders, and take action to remedy the disparities. The tool is grounded in research and was developed with input from a national expert panel. It has demonstrated utility in urban, rural, and suburban settings" (Prevention Institute, 2004).
Norris et al. (2008) community resilience model	An approach to the measurement of community resilience was proposed by Norris et al. in 2008. In a subsequent paper, Sherreib et al. (2010) combined this approach with publicly accessible population indicators, and applied it in a study of 21 counties of Mississippi. Their measure of community resilience, which is limited to the economic and social capacities of communities, indicated generally favorable correlation with archival and survey data.
Resilience Capacity Index (RCI)	The Resilience Capacity Index was developed by Kathryn A. Foster at the University at Buffalo Regional Institute. It provides "a single statistic summarizing a region's status on twelve factors hypothesized to influence the ability of a region to bounce back from a future unknown stress. The index permits comparisons across metropolitan regions and identification of strong and weak conditions relative to other metropolitan regions." Further details are available through the Building Resilient Regions project of the Institute of Governmental Studies, University of California, Berkeley (BRR, 2011).
Community Disaster Resilience Index (CDRI)	The CDRI was developed by a team of researchers at Texas A&M's Hazard Reduction and Recovery Center with support from NOAA. The quantitatively based metric uses the four phases of the disaster management cycle (preparedness, response, recovery, mitigation) and combines these with community capital assets (social, economic, physical, human, and natural capital). From the initial 120 candidate indicators, 75 were used in the index. Using subindexes based on each community capital (excluding natural capital), scores were averaged by each of the four capital assets and then averaged to compute the CDRI. The CDRI was then computed for

	Gulf of Mexico coastal counties (Peacock, 2010).
Center for Risk and Economic Analysis of Terrorism Events Economic Resilience Index (CREATE-ERI)	This index uses dollar values as a common denominator and is measured in terms of direct and indirect business interruption losses (usually as gross domestic product, or GDP). It is defined in terms of the standard "loss-triangle," and includes static considerations of resilience through improved allocation of existing resources and dynamic considerations of optimal investment to hasten recovery and reconstruction. In essence, it is defined as avoided losses divided by maximum potential losses. A major application was to the economic impacts of the September 11 terrorist attacks. The New York Metropolitan Area economy and the U.S. economy as a whole exhibited remarkable resilience. Ninety-five percent of the businesses in the World Trade Center area were able to relocate. Business interruption losses were incurred during the period in which relocation took place. Application of the index indicated that 72 percent of potential business interruption losses were reduced by this relocation resilience tactic (Rose et al., 2009).

INTERNATIONAL EFFORTS TO MEASURE RESILIENCE

Importantly, the international community also has a great deal of interest in disaster resilience. Although some of the international focus is clearly on resilience of individual countries, many of the international resilience initiatives focus on how to build a world more resilient to disasters. As a complement to the previous section on U.S. resilience models and metrics, this section reviews examples of resilience measurement at the global scale; one approach is described in detail, followed by a table summarizing other efforts.

International Strategy for Disaster Reduction

Following the International Decade for Natural Disaster Reduction (IDNDR) in the 1990s, the International Strategy for Disaster Reduction (ISDR) was developed through a gathering of stakeholders committed to reducing disaster risk and building the resilience of communities and actions. A major action plan was proposed in 2005 in Kyoto, Japan, known as the Hyogo Framework of Action (UNISDR, 2007) for the decade ranging between 2005 and 2015. Adoption of the plan by 168 states at that time was driven by the impact of the recent 2004 tsunami. The most recent review of the progress toward reaching that action plan was held in Geneva in May 2011 (UNISDR, 2011).

The action plan recognizes that the local community level is where the impact of disasters is most felt and where risk reduction is more needed—and that not addressing resilience may threaten nations' and communities' development gains. The plan also recognizes a need for international collaboration between various stakeholders interested in disaster risk reduction, namely states, regional organizations and institutions, international organizations, civil society, the scientific community, and the private sector.

The 2005 Hyogo Framework for Action (HFA) consists of five well-defined priorities for action (UNISDR, 2010):

- HFA-1, making risk reduction a national and local priority, with a strong institutional basis for implementation;
- HFA-2, identifying, assessing, and monitoring disaster risks, and enhancing early warning;
- HFA-3, building a culture of safety and resilience using knowledge, innovation, and education at all levels;
- HFA-4, reducing the risk in key sectors; and
- HFA-5, strengthening disaster preparedness for effective response at all levels.

Each priority for action is divided into several specific tasks. Each task is assigned specific measurable indicators, a method of monitoring progress, guiding questions, and specific tools to reach the desired level of disaster risk reduction (DRR). Further, implementation of each task is illustrated by one or several international case studies. Table 4.3 shows an example for HFA-1.

Table 4.3 Tasks Defined by HFA to Make Risk Reduction a National and Local Priority—HFA Priority 1

Making Risk Reduction a National and Local/City Priority with a Strong Institutional Basis for Implementation				
HFA Tasks	Local Indicators	National HFA Monitor Indicators	Guiding Questions	Tools
Task – 1 Engage in multistakeholder dialogue to establish foundations for disaster risk reduction (DRR)	• A local/city multisectoral platform for disaster risk reduction is functioning • Political commitment	A. National multisectoral platform for disaster risk reduction is operational	• Are different stakeholders engages in a continuing dialogue for disaster risk reduction? • Is there political consensus on importance of DRR? • What is the degree of participation of civil society in DRR? • Is local/city government supportive to a community vision for DRR?	• Multistakeholder dialogues; management information system
Task – 2 Create or strengthen mechanisms for systematic coordination for DRR	• Community participation and decentralized functions are ensured throughout the local	B. Community participation and decentralization are ensured through the delegation of authority and resources to local	• Are community participation and decentralization ensured through the delegation of authority and resources to the local/city level? • Is there official policy and strategy to support community-based disaster risk management in the city? • Are communities empowered to participate	• Stakeholder engagement mechanisms; local platform for DRR

	authority	levels	in disaster risk reduction? • Are city offices aware of their respective roles in reduction? • Are there committed and effective community outreach activities (DRR and related services, e.g. healthcare?)	
Task – 3 Assess and develop the institutional basis for disaster risk reduction	• Policy instruments and tools to support national institutional and legal frameworks • Legal and regulatory system	C. A legal framework for disaster risk reduction exists with explicit responsibilities defined for all levels of government. D. A national policy framework for disaster risk reduction exists that requires plans and activities at all administrative levels, from national to local levels	• Is responsibility for DRR planning and implementation devolved to city government and communities? • Are city government and communities equipped with human, financial, and organizational capacities/resources? • Are city government DRR policies, strategies, and implementation plans in place? • Are there relevant and enabling legislation (ordinance), land use regulations, building codes, etc. addressing and supporting DRR at the local level? • Are thre mechanisms for compliance and enforcement of laws, regulations, building codes, etc., and penalties for non-compliance defined by laws and regulations? • Is DRR integrated into planning at the local/city level in key sectors such as	• Development plan; land use plan; physical plan • Budget allocation for DRR • Disaster management ordinance; building code; fire code; zoning ordinance • Specific ordinances

			agriculture, climate change, education, environment, health, housing, poverty alleviation, and social welfare? • Are the roles and responsibilities for disaster risk reduction clearly designated? • Is the legal and regulatory system underpinned by guarantees of relevant rights to safety, to equitable assistance, to be listened to and consulted?	
Task – 4 Prioritize DRR and allocate appropriate resources	• Dedicated and adequate resources are available to implement DRR activities within the local authority	E. Dedicated and adequate resources are available to implement DRR plans at all administrative levels	• Are there institutional capacities for DRR at the local/city level? • Is budget allocated to local/city covernment and other local institutions adequate to enable DRR to be integrated into planning and actual activities? • Are financial resources available to build partnerships with civil society for DRR? • Are there logistical, and other such resources allocated for DRR? • Does the government provide training in DRR to local/city officials and community leaders? • Is a system of accountability in place, including transparency in the conduct of DRR and use of funds?	• Disaster risk management office; disaster coordinating council

Source: R. Shaw and Y. Matasuoko, UNISDR

Other International Resilience Metrics and Indicators

Other international metrics and indicators for vulnerability, risk, and resilience have also been developed. Table 4.4 provides a brief summary of some of these.

Table 4.4 Selected Summary of International Metrics and Indicators for Vulnerability, Risk, and Resilience

United Nations Development Programme (UNDP) Disaster Risk Index (DRI)	The DRI, introduced in 2004, measures the *average risk of death* per country in three types of disasters (earthquakes, tropical cyclones, and floods). It is a measure of vulnerability to a specific hazard that also accounts for the role of sociotechnical-humanistic and environmental issues that could be correlated with death and may point toward causal processes of disaster risks. The key steps in determining the DRI for *a specific hazard* include calculation of physical exposure in terms of number of people exposed to a hazard event in a given year; calculation of relative vulnerability in terms of number of people killed to number of people exposed; and calculation of vulnerability indicators using 26 variables. Based on the value of the DRI, and for a given specific hazard, countries are ranked according to their degree of physical exposure, relative vulnerability, and degree of risk (UNDP, 2004; Peduzzi et al., 2009).
Inter-American Development Bank Disaster Deficit Index (DDI)	The DDI, introduced in 2005, is an indicator of a country's economic vulnerability to disaster. It is limited to Latin America and the Caribbean. DDI is a measure of the likely economic loss related to a disaster in a given time period and for the economic coping capacity of the country (IDB, 2007).
Inter-Agency Standing Committee (IASC) In-Country Team Self-Assessment Tool for Natural Disaster Response Preparedness	Established in 1994, the IASC was created to be the primary mechanism for interagency coordination of humanitarian assistance at the international level. It is composed of representatives of all 14 leading UN agencies, non-UN humanitarian agencies, and three consortia of nongovernmental organizations. The In-Country Team Self-Assessment Tool for Natural Disaster Response Preparedness consists of a support chart and a checklist of issues and questions to self-assess the level of

	international standards. It also provides resources to address key concerns and propriety areas for disaster preparedness and response. See http://www.humanitarianinfo.org/iasc/.
United Nations University Institute for Environment and Human Security, World Risk Index	The World Risk Index, introduced in 2011 (UNU, 2011), indicates the probability that a country or region will be affected by an extreme natural event (earthquakes, storms, floods, droughts, and sea-level rise). It also focuses on (i) the vulnerability of the population (levels of poverty, education, food security, infrastructure, economic framework) to natural hazards, (ii) its capacity to cope with severe and immediate disasters as a function of governance, disaster preparedness, early warning systems, medical services, and social and economic security, and (iii) its adaptive precautionary measures against anticipated future natural disasters. The World Risk Index is also combined with local and project risk indexes.

THE COMMITTEE'S PERSPECTIVE

The preceding two sections have presented representative approaches to the measurement of resilience. They vary on many dimensions: top-down prescriptions versus community-based consensus; universal or adaptable, based on available data or requiring extensive data gathering; place-based or spatial, and focused on specific hazards and vulnerabilities or extensible depending on the context. This section introduces the committee's perspective, comments on each of these dimensions as they might apply to the committee's charge, and then moves to a discussion of the implementation of metrics.

First, the committee visited three different areas—New Orleans and the Mississippi Gulf Coast, Iowa, and Southern California—and recognized the degree to which community concerns vary. New Orleans was recovering from a major storm event and Iowa from a major flood event, whereas Southern California has a history of disastrous wildfires and landslides and must prepare for a future major earthquake event. In the committee's view, therefore, any approach to measuring resilience has to address multiple hazards, and has to be adaptable to the needs of specific communities and the hazards they face. By contrast, the SPUR model (see earlier section) concerns only earthquake hazard, though it could perhaps be generalized to other hazards.

Second, the committee met with communities of many sizes, from those in the greater metropolitan areas of Southern California to the small towns of the Mississippi Gulf Coast. It is clear that any approach the committee recommends must be place-based rather than spatial, in the meaning of those terms defined at the start of the chapter, and capable of dealing with a range of community sizes. Moreover some communities, such as the Lower Ninth Ward of New Orleans, will be very different in structure, spatial extent, and level of social organization than others. Again, the emphasis in the committee's approach to measuring resilience is on adaptability. This concern for community, place, and adaptability argues against any universal solution, such as that represented by the Argonne National Laboratory Resilience Index.

Third, the committee recognizes that many dimensions must contribute to an index, from the physical resilience of the built and natural environment and critical infrastructure to aspects of human/social resilience such as the existence of strong social networks, a strong economic base, or good governance. The examples that yield a single index—SoVI®, BRIC, and the Argonne National Laboratory Resilience Index—all focus on a single dimension, social vulnerability in the first case, community resilience in the second, and critical infrastructure in the third. SoVI®'s reliance on available Census data suggests that it would be difficult to extend its approach to other dimensions, while the Argonne approach requires substantial investment in data gathering, compared with the community-based data gathering of the Coastal Resilience Index, for example.

KNOWLEDGE AND DATA NEEDS

As mentioned in Chapter 3, the issues of data availability are critical not only for hazard and disaster informatics, but resilience metrics as well. However, it is not just data that constrain our ability to measure resilience. Better understanding on how to implement such a measurement system is also needed. What should be measured over what time frame and geographic scale? Should resilience be reassessed on a regular schedule, or should certain factors trigger a reassessment? Should scales be prescribed and uniform, or should they be adapted to meet specific circumstances? How should these indicators be measured (e.g., qualitatively, quantitatively)? Should these data be included into a single composite index or some other structure, and if a single index, how should the various components be weighted? By what means can it be determined that the right elements for the resilience index have been captured? How is the sensitivity of the index assessed? Addressing these issues through an integrated research program would assist the nation in providing the scientific backing for the development of a national resilience scorecard.

Moreover, such a research program could provide useful insights by making a systematic comparison of the different metrics proposed in the literature. Besides addressing the questions raised earlier in this paragraph, it would be very useful to compare metrics on the basis of cost, and the time and effort needed to implement and evaluate them.

SUMMARY AND RECOMMENDATION: IMPLEMENTING A MEASUREMENT SYSTEM

With this background, we now turn to the committee's conclusions and specific recommendations regarding metrics and indicators. Related topics have been discussed at several points in the report, including Chapter 3, where we discuss the lack of consistent, reliable data on the impacts of hazards and disasters that might feed into the measurement of resilience.

This chapter has focused on the importance of metrics and indicators that can be used to evaluate resilience, to provide baselines for comparison and the foundation for a system of tracking improvements. In essence, the committee concludes from the evidence gathered that **without some numerical basis for assessing resilience it would be impossible to monitor changes or show that community resilience has improved. At present, no consistent basis for such measurement exists. We recommend therefore that a National Resilience Scorecard be established.**

Until a community experiences a disaster and has to respond to and recover from it, demonstrating the complexity, volume of issues, conflicts, and lack of ownership are difficult. A national resilience scorecard, from which communities can then develop their own, tailored scorecards, will make it easier for communities to see the issues they will face without being subjected to the event and can support necessary work in anticipation of an appropriate resilience-building strategy. A scorecard will also allow communities to ask the right questions in advance.

In the preceding sections the committee's vision of such a scorecard was outlined. It should be readily adaptable to the needs of communities and levels of government, focusing specifically on the hazards that threaten each community. It should align with community goals and vision. It should not attempt unreasonable precision, either in the ways in which individual factors are measured, or in the ways they are combined into composite indicators. Rather, the scorecard should follow the examples presented earlier where qualitative and quantitative measures are mingled, and reduced where appropriate to ordinal (rankings) rather than interval or ratio scales.

The various indicators reviewed in this chapter vary greatly depending on the dimensions they assess, the sources of data they employ, and the ways in which they combine data to obtain indicators. However, certain commonalities emerge and provide useful guidance in the development of a Scorecard. While maintaining its commitment to local solutions and not wishing to be overly prescriptive, the committee emphasizes that it is imperative to include certain dimensions in the Scorecard:

- Indicators of the ability of critical infrastructure to recover rapidly from impacts (see, e.g., Section 4.2.1);
- Social factors that enhance or limit a community's ability to recover, including social capital, language, and socioeconomic status, and the availability of a workforce with skills relevant to recovery (see, e.g., Section 4.2.3);
- Indicators of the ability of buildings and other structures to withstand the physical and ecological impacts of disasters (e.g., ground shaking, severe wind and precipitation, inundation, fires (see, e.g., Section 4.2.5); and
- Factors that capture the special needs of individuals and groups, related to minority status, mobility, or health status (see, e.g., the T*H*R*I*V*E model in Section 4.2.6).

Although such a scorecard would be used as a self-assessment tool employed by individual communities, some central coordination and direction for the development of the scorecard is appropriate from the federal level. The committee concludes that responsibility for coordinating the development of a scorecard should rest with a single federal agency but be compiled through a national effort that engages with individuals and communities at all levels. The Department of Homeland Security appears to be the most appropriate agency for coordinating this collective endeavor. In summary, the committee concludes its work in the area of metrics and indicators with this recommendation:

***Recommendation.* The Department of Homeland Security in conjunction with other federal agencies, state and local partners, and professional groups should develop a National Resilience Scorecard.**

REFERENCES

Birkmann, J., ed. 2006. *Measuring Vulnerability to Natural Hazards.* New Delhi, India: TERI Press.

BRR (Building Resilient Regions). 2011. Resilience Capacity Index. Available at: http://brr.berkeley.edu/rci/.

CARRI (Community and Regional Resilience Initiative). 2011. Community Resilience System Initiative (CRSI) Steering Committee Final Report: A Roadmap to Increased Community Resilience. Available at: http://www.resilientus.org/library/CRSI_Final_Report-1_1314792521.pdf.

Cutter, S. L., and C. Finch. 2008. Temporal and spatial changes in social vulnerability to natural hazards. *Proceedings of the National Academy of Sciences of the United States of America* 105(7):2301-2306.

Cutter, S. L., B. J. Boruff, and W. L. Shirley. 2003. Social vulnerability to environmental hazards. *Social Science Quarterly* 84(2):242-261.

Cutter, S. L., C. G. Burton, and C. T. Emrich. 2010. Disaster resilience indicators for benchmarking baseline conditions. *Journal of Homeland Security and Emergency Management* 7(1). Available at http://www.bepress.com/jhsem/vol7/iss1/51.

Emmer, R., L. Swann, M. Schneider, S. Sempier, T. Sempier, and T. Sanchez. 2008. Coastal Resilience Index: A Community Self-Assessment. A Guide to Examining How Prepared Your Community Is for a Disaster. NOAA Publ. No. MAS GP-08-014. Available at http://research.fit.edu/sealevelriselibrary/documents/doc_mgr/434/Gulf_Coast_Coastal_Resilience_Index_-_SeaGrant.pdf.

FEMA (Federal Emergency Management Administration). 2001. Understanding Your Risks: Identifying Hazards and Estimating Losses. FEMA Publ. No. 386-2. Available at http://www.fema.gov/library/viewRecord.do?id=1880.

Fisher, R. E., G. W. Bassett, W. A. Buehring, M. J. Collins, D. C. Dickinson, L. K. Easton, R. A. Haffenden, N. E. Hussar, M. S. Klett, M. A. Lawlor, D. J. Miller, F. D. Petit, S. M. Peyton, K. E. Wallace, R. G. Whitfield, and J. P. Peerenboom. 2010. Constructing a Resilience Index for the Enhanced Critical Infrastructure Program. Argonne National Laboratory/Department of Energy Report No. ANL/DIS 10-9. Available at www.ipd.anl.gov/anlpubs/2010/09/67823.pdf.

Heinz Center (H. John Heinz III Center for Science, Economics, and the Environment). 2002. *Human Links to Coastal to Coastal Disasters*, Washington, DC: Heinz Center.

IDB (Inter-American Development Bank). 2007. Indicators of Disaster Risk and Risk Management: Program for Latin America and the Caribbean. Available at http://www.iadb.org/exr/disaster/ddi50.cfm.

Keeney, R. L., and H. Raiffa. 1976. *Decisions with Multiple Objectives: Preferences and Value Tradeoffs*. New York: Wiley.

Longley, P. A., M. F. Goodchild, D. J. Maguire, and D. W. Rhind. 2011. *Geographical Information Systems and Science*, 3rd Ed. Hoboken, NJ: Wiley.

Malczewski, J. 2010. *Multicriteria Decision Analysis in Geographic Information Science*. Berlin: Springer.

Massam, B. H. 1993. *The Right Place: Shared Responsibility and the Location of Public Facilities*. Harlow, UK: Longman.

Mileti, D. 1999. *Disasters by Design: A Reassessment of Natural Hazards in the United States.* Washington, DC: Joseph Henry Press.

Norris, F. H., S. P. Stevens, B. Pfefferbaum, K. F. Wyche, and R. L. Pfefferbaum. 2008. Community resilience as a metaphor, theory, set of capacities, and strategy for disaster readiness. *American Journal of Community Psychology* 41:127-150.

NRC (National Research Council). 2006. *Facing Hazards and Disasters: Understanding Human Dimensions*. Washington, DC: The National Academies Press.

Peacock, W. G., ed. 2010. *Advancing the Resilience of Coastal Localities: Developing, Implementing and Sustaining the Use of Coastal Resilience Indicators: A Final Report.* Hazard Reduction and Recovery Center, Texas A&M University. Available at http://archone.tamu.edu/hrrc/Publications/researchreports/downloads/10-02R_final_report_grant_NA07NOS4730147_with_cover.pdf.

Peduzzi, P., H. Dao, C. Herold, and F. Mouton. 2009. Assessing global exposure and vulnerability towards natural hazards: The Disaster Risk Index. *Natural Hazards and Earth System Sciences* 9:1149-1159.

Phillips, B. D., D. S. K. Thomas, A. Fothergill, and L. Blinn-Pike, eds. 2010. *Social Vulnerability to Disasters.* Boca Raton, FL: CRC Press.

Prevention Institute. 2004. *A Community Approach to Address Health Disparities: T*H*R*I*V*E: Toolkit for Health and Resilience in Vulnerable Environments.* Available at http://minorityhealth.hhs.gov/assets/pdf/checked/THRIVE_FinalProjectReport_093004.pdf.

Rose, A., G. Oladosu, B. Lee, and G. Beeler-Asay. 2009. The economic impacts of the 2001 terrorist attacks on the World Trade Center: A computable general equilibrium analysis. *Peace Economics, Peace Science, and Public Policy* 15(2), Article 4.

Saaty, T. L. 1988. *Decision Making for Leaders: The Analytical Hierarchy Process for Decisions in a Complex World.* Pittsburgh, PA: University of Pittsburgh Press.

Sherrieb, K., F. Norris, and S. Galea. 2010. Measuring capacities for community resilience. *Social Indicators Research* 99(2):227-247.

SPUR (San Francisco Planning and Urban Research Association). 2008. *Defining What San Francisco Needs from Its Seismic Mitigation Policies.* Available at http://www.spur.org/publications/library/report/defining-what-san-francisco-needs-its-seismic-mitigation-policies#disaster.

START (National Consortium for the Study of Terrorism and Responses to Terrorism). 2011. *Developing Community Resilience for Children and Families.* Available at http://www.start.umd.edu/start/research/investigators/project.asp?id=30.

Tuan, Y.-F., 2007. *Space and Place: The Perspective of Experience.* Minneapolis: University of Minnesota Press.

UNDP (United Nations Development Program). 2004. Reducing Disaster Risk: A Challenge for Development. Available at http://www.grid.unep.ch/activities/earlywarning/DRI/.

UNISDR (United Nations International Strategy for Disaster Reduction). 2007. Hyogo Framework for Action 2005-2015: Building the Resilience of Nations and Communities to Disaster. Available at http://www.unisdr.org/files/1037_hyogoframeworkforactionenglish.pdf.

UNISDR. 2010. A Guide for Implementing the Hyogo Framework for Action by Local Stakeholders. Available at http://www.unisdr.org/files/13101_ImplementingtheHFA.pdf.

UNISDR. 2011. Themes and Issues in Disaster Risk Reduction. Available at http://www.preventionweb.net/files/19646_themesandissuesindrrwithdefinitions.pdf.

UNU (United Nations University). 2011. *World Risk Report 2011.* Berlin, Germany: Alliance Development Works. Available at http://www.ehs.unu.edu/file/get/9018.

Characteristics of a resilient community include "a sense of ownership and responsibility towards the community . . . (and) the ability to communicate and share resources during . . . or very shortly after . . . the disaster."
Iowa citizen, 2011

5

Building Local Capacity and Accelerating Progress: Resilience from the Bottom Up

National resilience emerges, in large part, from the ability of local communities to plan and prepare for, absorb, respond, and recover from disasters and adapt to new and diverse conditions such as economic growth and decline, technology innovations, and rising sea level. Interventions to enhance resilience to disasters require both the "bottom-up" approaches at the local community level detailed in this chapter and the "top-down" strategies at the federal and state levels addressed in Chapter 6.

Bottom-up interventions are essential because local conditions vary greatly across the country and often jurisdictional issues exist around who can respond to the call to increase resilience, and when. The nation's communities are unique in their history, geography, demography, culture, economic enterprise, governance, and infrastructure. Moreover, the risks faced by every community vary according to local hazards and exposure levels, vulnerabilities, and capacities to mitigate. Plans to enhance resilience to hazards and disasters in one locale may not match community baselines, assets, and requirements in another (see Chapters 2 and 3; NRC, 2011b). Building resilience in the face of disaster risk can also have benefits for a community even in the absence of a disaster in advancing the social capital for dealing with more mundane community challenges.

Although each community is responsible for developing its own path toward greater resilience, the committee identified some universal steps that can aid local communities in making progress to increase their capacity to withstand and recover from disasters. These steps are intended to strengthen both the social infrastructure, which reflects the ties among people and their commitments to collective problem solving, and the physical infrastructure, which includes the built environment and critical lifelines that house and sustain human activity. These steps include

- Engaging the whole community in disaster policy making and planning;
- Linking public and private infrastructure performance and interests to resilience goals;

- Improving public and private infrastructure and essential services (such as health and education);
- Communicating risks, connecting community networks, and promoting a culture of resilience;
- Organizing communities, neighborhood, and families to prepare for disasters;
- Adopting sound land-use planning practices; and
- Adopting and enforcing building codes and standards appropriate to existing hazards.

This chapter reviews the essential elements of these steps as a means for communities to secure a foundation either to begin, or to help reinforce, initiatives and programs to enhance resilience.

WHOLE COMMUNITY ENGAGEMENT

Consensus is emerging among policy makers (DHHS, 2009; DHS, 2010; FEMA, 2010, 2011), practitioners (Patton, 2007; Waugh and Streib, 2006), and researchers (NRC, 2010, 2011b) that collaboration between the private and public sectors can enhance the disaster resilience of a community. Indeed, the National Research Council has released a number of recent reports that spotlight the role of private–public partnerships and collaborative organizational structures in strengthening community resilience to disasters (NRC, 2005a, 2006a, 2009, 2010, 2011b).

The most pressing issue in moving forward with this kind of collaboration is how to involve the community and businesses—both part of the private sector—effectively and productively in decision making and capacity building for disaster resilience. During the course of this study, the committee has identified four mechanisms for engagement that could assist communities in building capacity and becoming an effective part of the decision making process for disaster resilience (Table 5.1). These mechanisms tie back to the risk management cycle outlined in Chapter 2.

Table 5.1 Mechanisms for Community Engagement in Disaster Policy Making

Mechanism	Purpose
Development of broad-based community coalitions	Rather than just an instrument to secure a community's concrete commitment to disaster resilience, the development of a broad-based community coalition is itself a resilience-generating mechanism in that it links people together to solve problems and builds trust.
Involvement from a diverse set of community members—the "full fabric" of the community	Because no single entity can deliver the complete public good of resilience (see Chapter 3), resilience becomes a shared

	value and responsibility. Collaboration in fostering interest in resilience in the community can ensure that the full fabric of the community has the opportunity to be included in the problem-solving endeavor—and that it represents public and private interests and people with diverse social and economic backgrounds.
Building organizational capacity and leadership	Meaningful private–public partnerships for community resilience depend upon strong governance and organizational structures, leadership, and sustained resources for success.
Resilience plan	A priority activity for a local disaster collaborative is planning for stepwise improvements in community resilience.

Community Coalitions to Foster Community Resilience

Teaming up to take proactive steps to manage risks—such as a resilience private–public coalition—embodies several preconditions for successful adaptation by a community facing a major disturbance or stress. In their interdisciplinary review of the resilience literature, Norris et al. (2008) conclude that those communities that adapt well to adversity—and quickly return to a state of population wellness—do so through reliance on four key resources and their interactions: (1) economic resources (including the level and diversity of, and access to, these resources), (2) social capital (including organizational and interpersonal links, the sense of community among the citizens, and citizens' own participation in community life), (3) information and communication (which have to involve trusted information sources and outlets), and (4) community competence (group skills for collective action and a system of shared beliefs). Another leading model of resilience similarly recognizes resources, communication, connectedness, commitment, and shared values, and critical reflection and skill building as major contributing factors to a community's ability to rebound from disasters (Pfefferbaum et al., 2008).

In this context, private–public partnerships become an essential vehicle for enhancing community resilience to disasters (e.g., the Safeguard Iowa Partnership; see NRC, 2011b). Such partnerships have the potential to focus diverse social networks around a common cause, to facilitate the sharing of information essential to understanding risk and means to reduce it, and to apply the intellectual strengths of many people to the problems of building resilience to disasters. These partnerships serve as coalitions to act as a collective and cohesive unit that can define, address, and solve problems for the betterment of the community (Pfefferbaum et al., 2008). Experience in the emergency management sector illustrates how private–public coalitions are integral to community efforts to build resilience (Box 5.1).

BOX 5.1
Emergency Management and Unity of Effort to Increase Resilience

The following is extracted from the document "Principles of Emergency Management" (IAEM, 2007) and identifies some of the principles of emergency management that relate to the role of emergency managers as practitioners of risk management.

"Emergency managers ensure unity of effort among all levels of government and all elements of a community. In the early 1980s, emergency managers adopted the Integrated Emergency Management System (IEMS), an all-hazards approach to the direction, control and coordination of disasters regardless of their location, size and complexity. IEMS integrates partnerships that include all stakeholders in the community's decision-making processes. IEMS is intended to create an organizational culture that is critical to achieving unity of effort between governments, key community partners, non-governmental organizations (NGOs) and the private sector.

Unity of effort is dependent on both vertical and horizontal integration. This means that at the local level, emergency programs have to be integrated with other activities of government. For example, department emergency plans have to be synchronized with and support the overall emergency operations plan for the community. In addition, plans at all levels of local government ultimately have to be integrated with and support the community's vision and be consistent with its values.

Similarly, private sector continuity plans have to take into account the community's emergency operations plan. Businesses today are demanding greater interface with government to understand how to react to events that threaten business survival. Additionally, businesses can provide significant resources during disasters and thus may be a critical component of the community's emergency operations plan. In addition, given the high percentage of critical infrastructure owned by the private sector, failure to include businesses in emergency programs could have grave consequences for the community.

In this sense of using coalitions to best advantage to increase disaster resilience, local emergency management programs also have to be aligned and synchronized with higher-level plans and programs in government. The need for this kind of synchronization is most noticeable in the dependence of local government on county, state and federal resources during a disaster [see below; also Chapter 6]. If plans have not been aligned and synchronized, allocation of resources may be delayed.

Integrating emergency management into daily decisions in the community is important so that critical decisions are not made only during times of disasters. While protecting the population is a primary responsibility of government, this kind of protection is difficult to accomplish without building partnerships among disciplines and across all community sectors, including the private sector and primary communications entities such as the media."

The Full Fabric of Community Woven into Resilience Coalitions

Resilience is a shared responsibility. As outlined in Chapter 3, responsibility for strengthening resilience does not rest solely with government, particularly given the wealth of resources and capacities resident in the community itself. In the United States, the public sector constitutes just 10 percent of the total workforce (NRC, 2011b). The remaining 90 percent works in both the private sector—from small, individually owned businesses to national and global conglomerates—and in nongovernmental organizations (NGOs) and faith-based organizations (FBOs). Ownership, management, and intimate technical understanding of the country's critical infrastructure—water, power, communication, health care, and transportation networks—rests largely in private hands. Community- and faith-based groups usually have established leadership and communication structures and social standing in the community. They have proven powerful allies in disaster response and recovery (Wachtendorf and Kendra, 2004) and thus have natural roles in the building of overall disaster resilience (Box 5.2). Often, they are assisted by their networks outside of the disaster region, thus improving the response to the disaster, and providing valuable experience for groups in other regions. For example, in the case of Hurricane Katrina, churches around the country assisted their counterparts in New Orleans and Mississippi. Universities did the same, taking in students from the affected region for the fall semester, often at no charge.

BOX 5.2
Health Department Uses Community Approach to Protect People Against Carbon Monoxide Poisoning

In December of 2006, record-setting torrential rains and high wind speeds in King County, Washington, interrupted power to 1.5 million utility customers. As power outages wore on, area hospitals saw unprecedented numbers of patients with carbon monoxide (CO) poisoning. This health threat accounted for 8 of the state's 15 storm-related fatalities. The profile of early patients showing up at local hospitals with evidence of CO poisoning suggested that immigrant groups were at increased risk. Faced with no power, for instance, some Somali and Vietnamese immigrants turned to cooking and warming themselves over charcoal grills indoors. The difficulties conducting effective outreach to immigrant and refugee communities during this power outage propelled Public Health–Seattle and King County to reevaluate communications procedures to include the whole of the community.

Working with their Vulnerable Populations Action Team (VPAT), the health department developed a Community Communications Network consisting of over 150 community organizations to relay information to the people they serve. Stronger relationships developed with many of these organizations, leading to the formation of new groups who were ready to mobilize, such as a Somali Health Board of ethnic community leaders. Informational interviews and focus groups with diverse members of the local communities lead to better information about trusted sources of information and effective methods of distribution.

In January 2012, the region experienced a snow and ice storm that led to a similar power outage situation. However, with the strengthened resilience coalition in place, Public Health–Seattle and King County rapidly disseminated CO information to community partners using channels recommended by the community. Flyers in 25 languages blanketed hardware stores, grocery stores, language schools, apartments and businesses in identified neighborhoods. Information was broadcast over ethnic media outlets, community webcasts, loudspeakers at Lunar New Year festivals, taxicab dispatchers, and through a robo-call from a local mosque. Most importantly, hundreds of community partners received CO warnings and relayed information to their constituents. As a result, the number of CO poisonings was a tenth of what they were 5 years prior, and there were no fatalities. This culturally sensitive, social network-driven response likely reduced poisoning incidents. At the same time, it built up relationships and goodwill between the health department and diverse community segments.

Sources: Broom (2007); Public Health–Seattle and King County (2006, 2012a,b).

Successful collaborations in the interest of resilience also require input from people representing the full spectrum of a community's members including minorities, the disenfranchised, those with disabilities, children, senior citizens, and other subgroups that are potentially vulnerable to disaster impacts. Integrating the perspectives and contributions of these populations into resilience-enhancing activities is especially important because the chances for greater victimization during a disaster are unevenly distributed in society, as are opportunities for enhanced safety (Tierney et al., 2001; NRC, 2006b; Enarson, 2007; Morrow, 2008; Mary Claire Landry, personal communication, 2011 [see also Appendix B and NRC, 2011a]). At the same time, the resilience of at-risk populations and the perspective that they can bring to disaster risk reduction cannot be underestimated (Schoch-Spana et al., 2008). People who have coped with daily disasters such as poverty, deprived neighborhoods, or high rates of crime and violence may not see themselves as vulnerable, and ethnic groups cut off from mainstream society may still have strong internal ties that protect against some disaster impacts. An example is the Vietnamese community in

New Orleans and their recovery after Hurricane Katrina (Box 5.3; see also NRC, 2011a).

At the broadest scale of the nation, integrating the full fabric of a community into a resilience-enhancing collaboration may require a diverse set of strategies and incentives to motivate participation. People may be more inclined to embrace disaster loss reduction and enhanced public safety when they see something of personal value in reaching for these goals (Geis, 2000). A commercial enterprise, for example, may be motivated to engage in resilience-enhancing initiatives by the potential return on investments (e.g., reduced chances for business interruption), by access to information that improves business continuity planning, and by an increase in its public standing in the community (NRC, 2011b). A good example of this occurred in Rutland, Vermont, which was severely affected by flooding from Hurricane/Tropical Storm Irene in August 2011, as was the surrounding region. The only large grocery store in the area was badly flooded, but a very functional, temporary solution was established to allow residents to meet their daily needs and return to a sense of normalcy (Figure 5.1).

BOX 5.3
Seeing Itself as Self-Reliant, a Vietnamese Community Weathers Serial Disasters in the Gulf

About 8,000 of the approximately 40,000 Vietnamese residents on the U.S. Gulf Coast live in New Orleans East, among a large African American and Hispanic population (NRC, 2011a). Many community members came from Vietnam in 1975, when a large number of South Vietnamese immigrants arrived in the United States. Presently, the East New Orleans community now includes the children and grandchildren of these original immigrants. The residents with whom the committee spoke during their visit to the area described their relative isolation before Katrina as one without interaction with other sociocultural groups living in the area, but that all of these groups joined together after Katrina. They described themselves as self-reliant people who had built new lives after fleeing Vietnam. Community members spoke of their collective efforts to get everyone to safety during and immediately after Hurricane Katrina in a community where they said ~30 percent were elderly. The pastor of the local Catholic church where many of the residents attend services, Rev. Vien The Nguyen, took a boat through flooded neighborhoods to check on community members; they lost only one elderly person to the storm out of the entire population. Their evacuation planning was coordinated through the church and the local radio station directly through community initiatives.

Because fishing was a main source of income, Hurricane Katrina significantly affected a large segment of the community's livelihood, and after the storm, the community collectively decided to work together to rebuild, sharing with the community building and carpentry skills that some community members had developed back in Vietnam. Of the experience, one community member said "We are all carpenters now (NRC, 2011a)." After repairing their houses, they helped each other repair their boats, without bank loans, and with little immediate help from federal or other government sources. Nonetheless, when some federal funding did arrive, the community members expressed some surprise and gratitude for the additional support.

As with other communities along the Gulf Coast, the Deepwater Horizon blowout and subsequent oil spill in 2010 affected the community in East New Orleans again. With one-third of the community in the fishing industry, the fishing season was severely affected and anticipated income from the fishing industry put into doubt.

The Vietnamese community members stressed their ability to plan as a community, to carry out their plans when disaster struck, to rebuild, and to work together to seek improvements in their community following both disasters. From an outside perspective, their refugee experience and cultural values around helping each other helped to build both resilience and a sense of community, which served as points of strength during natural and human-induced disasters.

In California's Alameda County, Collaborating Agencies Responding to Disaster (CARD) promotes disaster preparedness among grassroots groups and social services agencies serving vulnerable populations, by providing them with dual-use tools. CARD, for instance, has transformed the traditional Incident Command System into a leadership course that improves the skills of nonprofit organizations at managing resources and relating to other agencies on a day-to-day basis (Schoch-Spana et al., 2008).

FIGURE 5.1 Grocery store in a tent. This tent began operating shortly after the flooding in Vermont as a result of Hurricane Irene. A generator truck is off to the left and the brick and mortar store (the damaged grocery store) behind the tent. The makeshift tent supplied residents' needs through at least early January 2012. Source: Allan H. Stern.

Building a diverse constituency base around the public goal of disaster resilience has the added benefit of countering interests that otherwise motivate people to engage in risky behavior. Driven by a profit motive, for example, developers may elect to build homes in hazard-prone areas such as along the nation's coasts; similarly, people continue to purchase homes in these areas, driven by the wish to live in what they perceive as a desirable location. In addition, development of vulnerable coastal zones or river floodplains may be encouraged by local decision makers who see such development as an opportunity to expand the tax base for their jurisdiction. On the other hand, strategies exist both to deter people from either building or choosing to live in hazard-prone areas and to mitigate against existing hazards through specific building techniques and approaches (see structural and nonstructural measures in Chapter 2). A broad-based constituency may help build the local political will to execute community resilience-enhancing measures possible only through public institutions and government action. Positive examples include Tulsa, Oklahoma's land-use reforms and stormwater utility fees in support of the local flood control program (Meo et al., 2004), or locally supported taxes to subsidize the retrofitting of public buildings against seismic hazards, in the case of Berkeley, California (Chakos et al., 2002; see also Chapter 2). However, these kinds of systematic remedies in the public interest can be unpopular to some and prove difficult to establish more broadly in the country. Cedar Rapids, Iowa, while committed to a long-term recovery and mitigation strategy following the dramatic 2008 floods, is nonetheless challenged with how to cover its portion of the costs associated with a proposed flood protection and management system (Chuck Wieneke, personal communication, March 8, 2011).

Organizational Capacity and Leadership to Sustain Collaboration

Strong leadership and a sustained organizational base are critical for facilitating collaboration to enhance resilience. Successful community-based partnerships leading to improved hazard mitigation practices often have had key, inspired individuals or champions who have catalyzed larger institutional changes (Prater and Lindell, 2000). Such was the case in the Berkeley, California, and Tulsa, Oklahoma, cases mentioned earlier. Institutionalizing a shared vision improves the likelihood that the collaboration will be sustained even after the dynamic leadership changes (NRC, 2010).

Sustaining public–private resilience coalitions requires an individual or group dedicated to advancing the collective project and keeping resilience on the community's overall agenda when interest might otherwise lag or opposition is encountered. For example, local coordinators for government-sponsored programs such as FEMA's Project Impact, preparedness coordinators for local health departments, and dedicated staff and institutional champions have been suggested as key ingredients for successful collaborations for resilience-building activities (Roussos and Fawcett, 2000; Tierney, 2000; Avery and Zabriskie-Timmerman, 2009; Orians et al., 2009).

Although the coordinating function is seen as central to the longevity and effectiveness of a resilience-focused collaboration, opinions are divided as to whether government or a nonpartisan entity is the appropriate actor to fulfill this duty (NRC, 2011b). Whether a governmental entity or a nongovernmental group is the final accountable entity for integrating individuals, communities, and businesses to increase community resilience, any resilience-focused collaboration is necessarily a part of consistent support for the legal authority of emergency management agencies. Regardless of where responsibility for coordination lies, resource allocation for this management function is important.

A Resilient Future Relies upon a Commitment to Planning

Communities can greatly increase their resilience through short- and long-term planning that is developed, endorsed, and implemented by officials of government, business, health care, education, and community-based organizations (CBOs). The plan would include risk management (see Chapter 2), community organization with chartered roles and responsibilities, named leaders, and a jointly developed community-committed culture; a resource management function to assign value to the community assets (plans, programs, control/oversight; see Chapter 3); and metrics to assess progress (see Chapter 4) (Table 5.2).

To maximize effective implementation, a resilience plan may align its goals with a culture of self-reliance; community self-sufficiency; and mutual aid and interdependencies with neighboring communities, state and federal government entities, and NGOs CBOs, and FBOs. Although specific resilience goals may vary among communities, a common set of principles (see Chapter 1) may help build a culture of resilience and steps toward achieving higher levels of disaster resilience.

Table 5.2 Suggested Elements of a Local Resilience Plan

Program Element	**Attributes**
Community organization	Reflects community structure and leadership
Standards and codes	Represents current and needed building and development codes, standards, and zoning ordinances, where compliance and enforcement are emphasized
Performance metrics and resilience rating system	Represents assessment status and needs for essential progress in building resilience and desired performance of critical services and infrastructure following disruption

Education and communication	Represents critical education, outreach, and communication plans and practices for resilience to reach all community members
Local capacity	Designed to establish baselines and close essential capacity gaps in the community
Resource management	Integrates resources such as human and financial capital, mutual aid agreements, asset management strategies, essential relationships within interdependent communities and agencies

LINKING PRIVATE AND PUBLIC INFRASTRUCTURE INTERESTS

Lifelines

The second step for enhancing resilience at the local level is to link private and public infrastructure performance and interests. Accountability for critical infrastructure systems is dispersed across the public and private sectors (see Chapter 3). Lifelines— essential utility (e.g., domestic water/wastewater systems, industrial waste systems, power systems, fuel systems, telecommunications systems) and transportation systems (e.g., highways, bridges, railroads, transit systems, airports, seaports, waterways)—are both publicly and privately owned and share the attributes of being distributed systems, rather than isolated facilities. They also provide products and services that are transferred through networks that often cross legal and jurisdictional boundaries (ALA, 2005). To complicate matters, these lifelines are in variable states of age and condition. It is essential to conduct assessments of the quality and condition of these, and to make needed improvements in order to enhance resilience.

Genuine resilience of community lifelines cannot be achieved in piecemeal fashion by private and public entities acting on their own. Instead, as Chapter 3 outlined, resilience requires that local infrastructure leaders come together to assess the status, vulnerability, and interdependencies of their holdings; set performance metrics for individual components and entire systems; and develop plans for enhancing the infrastructure's ability to withstand failure and for speeding the resumption of operations during disaster response and recovery (Box 5.4). As a locally based method of risk management, public–private infrastructure coalitions can also run joint community exercises using stress scenarios to test their systems for weak spots, initiate operational improvements to keep their enterprises functioning, and establish multiyear regional capital investment priorities.

BOX 5.4
San Francisco "Lifelines Council" Strives for Earthquake Resilience Through Infrastructure Upgrades

On October 14, 2009, San Francisco held its first Lifelines Council meeting realizing a vision proposed by the San Francisco Planning and Urban Research Association (SPUR), a community-based nonprofit committed to civic planning that represents citizens' voices and nurtures a vital urban center. Recognizing disaster planning as essential to the city's well-being, SPUR launched the Seismic Hazard Mitigation Initiative[a] to advance greater understanding of what it would take—from an engineering standards perspective—for "the city to remain safe and usable after a major earthquake" (Poland, 2009, p. 4). If San Francisco hoped to rebound quickly and minimize disaster costs, then the city needed to take active steps toward measuring and improving the performance of local buildings, utility systems, and transportation networks under the stress of a major earthquake. A highly recommended step was the creation of a local "lifelines council" to engage infrastructure owners and operators in comprehensive planning for seismic mitigation (Barkley, 2009). Chaired by the mayor's office, the proposed council includes representatives of city agencies responsible for local lifeline sectors (e.g., San Francisco Public Utilities Commission, Municipal Transportation Authority) and city departments with a coordinating role (e.g., Public Works, Emergency Management); state, regional, and private-sector entities operating or regulating lifelines that serve the city (e.g., CalTrans, AT&T, Bay Area Rapid Transit); and risk and industry experts (Barkley, 2009). Among the council's charges were:

- Coordinating planning across sectors, given lifeline interdependence (e.g., electric power runs the water and wastewater systems);
- Developing and adopting common performance goals and standards;
- Guiding a seismic performance audit of lifelines in the city, thus providing an evidence base for the city to establish priorities for system improvements;
- Establishing a funding plan for modifications to city-owned systems and for assistance to other system owners for modifications in areas of overwhelming public interest; and
- Communicating to political leaders and the public the value of improved lifeline performance, enlisting their support for potential service costs to cover enhancements (Barkley, 2009).

[a]See http://www.spur.org/publications/library/report/lifelines.

Many reports and studies address the importance of protecting the nation's critical infrastructure (Flynn, 2008; Chang, 2009; National Infrastructure Advisory Council, 2009a,b, 2010) and improving its resilience. However, the majority of these studies focus on national strategies and policies (top-down strategies), rather than on more locally based options. Community-based research suggests benefits for communities by engaging in development of complementary strategies for linking private and public goals and interests for upgrading and hardening infrastructure such as constructing levees and restoring wetlands as flood control projects (Guikema, 2009), enhancing the seismic resilience of communities (Bruneau et al., 2003), or enhancing the resilience of major commerce and transportation systems such as at the Port of Los Angeles (Box 5.5).

BOX 5.5
The Nation's Busiest Port Merges Green and Resilience Goals

The Port of Los Angeles (below) is one of the nation's busiest ports, and together with the adjacent Port of Long Beach handles the largest volume of containerized freight of any port complex in the United States. In 2010 the Port handled over 540,000 TEUs (20-foot equivalent units; 40-foot containers count as two in this statistic). Containerized cargo is moved out of the Port on rail via the Alameda Corridor to the yards near Downtown Los Angeles, and by the approximately 12,000 trucks that operate in and out of both ports. The immense size of the port (over 7,500 acres of land and water) and the value and importance of the freight handled make for a very significant and demanding security mission. The potential impact of a disruption at the Port is immense, both within the Los Angeles Basin, with a population approaching 20 million, and across the United States.

The Port of Los Angeles is a public entity, but operates as a self-supporting business by taking profits and putting them back into maintaining and upgrading infrastructure. The Port is not self-sufficient, but relies on other infrastructure providers for water and power, and so enhancing resilience requires cooperation among different sectors, agencies, and jurisdictions. As part of its modernization and capital improvement plan, the port is committed to green growth principles: that is, it "will maximize its social, economic, and environmental objectives to find mutually reinforcing solutions, recognizing their interdependencies. Likewise, the social, economic, and environmental impacts of port actions are considered when assessing organizational performance" (Port of Los Angeles, 2011). One specific effort is implementing a green building policy in which all Port structures are built to LEED gold standards. In both rhetoric and practice, the Port of Los Angeles exemplifies locally based efforts to enhance resilience.

Source: Port of Los Angeles, 2011.

The Port of Los Angeles is a facility of critical importance with links to other major components of the infrastructure in Southern California (utility and water services, freeways, railroads). Source: Gerry Galloway.

Resilience to Disaster in the Health Arena

Other infrastructure in communities is affected in similar ways. For example, the U.S. health care system is a dispersed, mostly for-profit system in which individual hospitals and other institutions (e.g., clinics, nursing homes, dialysis centers) compete for patients and resources at the same time that governmental public health agencies are responsible for the well-being of entire populations (Toner et al., 2009). Unlike most countries, the United States has no national health system. Also, there is no universal access to health care, even preventative care such as immunizations. The Department of Health and Human Services, including agencies such as Centers for Disease Control and Prevention and local and state health departments have some responsibility for guidelines, coordination, and even regulation in emergencies, but responsibility for acting on these remains at local levels with wide variation in capacity. In addition, there is no national- or state-level system for housing medical records electronically in ways that would permit retrieval of essential individual health information in emergencies. This was clearly demonstrated in the Hurricane Katrina disaster (see also Chapter 3). A major problem for those evacuated before, during, and after Hurricane Katrina was the absence of medical records indicating major health problems and medications taken routinely. People fled with no or insufficient supplies of medication (NRC, 2011a). Also, their essential care was interrupted when health care facilities became impaired, they lost resources with which to pay for care even if available, and they were displaced from the usual sources of treatment and support (Kessler, 2007; Zoraster, 2010). LTG Russell Honore, Commander, Joint Task Force, Katrina. has argued, "The health of a community before any crisis has a direct correlation to the magnitude of the health crisis after the event" (Honore, 2008). Research and responder experience have borne this out repeatedly. For instance, Gulf residents saddled with the highest burden of chronic disease prior to the infamous 2005 hurricane (many of them poor and medically underserved) were the hardest hit, as noted above.

Individuals with chronic disease such as asthma, heart disease, and diabetes, too, were among those at highest risk for developing flu-related complications during the 2009 H1N1 influenza pandemic (CDC, 2012). At that time, racial and ethnic minorities were at a threefold disadvantage medically because they were at higher risk of being exposed to the H1N1 virus, of being susceptible to its complications (because of a high prevalence of chronic conditions and immunosuppression), and of having impaired access to timely and trusted health information, vaccination, and treatment (Quinn et al., 2011).

From a health perspective, resilience to disasters and catastrophic health events involving infectious disease is grounded in both a robust population and a robust public health preparedness system. Leading figures in U.S. public health and national security have spotlighted the importance of promoting healthy lifestyles, investing in preventive care, and reversing health disparities as key to increasing the country's overall resilience (Honore, 2008; Lurie, 2009; Satcher, 2011). At the same time, they have underscored the importance of building and sustaining a network of ready and responsive individuals and institutions poised to reduce morbidity and mortality levels should a major crisis emerge. These priorities are not being upheld by necessary resources.

Assuring access to preventive care, aggressively providing secondary prevention, and implementing population-level interventions to prevent chronic disease are important means of creating a robust and resilient population (Lurie, 2009). Remedying health inequities, too, will help build resilience and reduce the medical footprint of hazards, disasters, and epidemics (Kessler, 2007; Honore, 2008; Zoraster, 2010; Quinn et al., 2011; Satcher, 2011). Fundamental resilience—that embedded within the very health and wellness of the population—helps mitigate the potential medical consequences of a disaster or epidemic. So, too, does a capable and comprehensive public health emergency preparedness system. Strong health agencies at the state and local level, backed up with federal support, serve as the coordinating backbone for this system that also incorporates individuals, businesses, and civil society groups (IOM, 2008).

The importance of public health agencies was underlined in the measures that federal decision makers took to reinvigorate the U.S. public health infrastructure in the wake of the 9/11 terrorist attacks and anthrax letter crisis in 2001. The Public Health Security and Bioterrorism Preparedness and Response

Act of 2002 established a system of federal grants to state and local health departments to upgrade their readiness and response capabilities for bioterrorism and other public health emergencies.[1] From FY 2001 to FY 2012, an estimated $8.95 billion has been awarded to support state and local public health preparedness activities (Franco and Sell, 2012). This infusion of funds has drastically improved the country's ability to handle extreme health events (Nuzzo, 2009; CDC, 2011a,b; Trust for America's Health, 2011). All state health departments, for instance, have staff on call all day and every day to evaluate urgent disease reports (Nuzzo, 2009). In 1999, only 12 states had this capability. All 50 states and the District of Columbia now have staff trained in their roles and responsibilities during an emergency (Nuzzo, 2009). Again, in 1999, only 12 states had this capability.

State and local health departments continue to work hard at enhancing the full range of preparedness capabilities including biosurveillance, medical countermeasure dispensing, emergency operations coordination, emergency public information and warning, and medical surge management (CDC, 2011c). Measurable advances in public health preparedness over the last decade, however, are now in jeopardy because of declines in federal, state, and local government budgets, cuts in the public health workforce, and an evolving list of public health threats (Nuzzo, 2009; CDC, 2011a,b,c; Trust for America's Health, 2011). Projected pressures on public health by 2020 include an increase in the U.S. population from 308 million to 336 million, the demands of more diversified age groups (e.g., a 54 percent increase of citizens over 65) on an already overburdened health care system, and mass migrations due to extreme weather events (CDC, 2011a).

Community health networks are another example of linking private and public infrastructure interests at the local level to foster resilience. Over the past decade, health care coalitions have emerged as an adaptive mechanism to overcome differences between the individualized nature of health care delivery and the large-scale, population-based demands for care in a public health emergency (Courtney et al., 2009). As institutionalized entities, healthcare coalitions are more frequent now across the United States since the establishment in 2002 of the Hospital Preparedness Program (HPP, though variously named over the years), a federal grant initiative mandated by Congress to upgrade local healthcare readiness for biological attacks and other public health emergencies (HRSA, 2002). Though initially focused on enhancing the preparedness of individual hospitals for biological incidents, the program has evolved and expanded to encourage greater all-hazards coordination among healthcare facilities in the same community or region (Courtney et al., 2009).

---

[1] Public Health Security and Bioterrorism Preparedness and Response Act of 2002. Pub. L. No. 107-188, 107th Cong., June 12, 2002. Available at http://www.gpo.gov/fdsys/pkg/PLAW-107publ188/pdf/PLAW-107publ188.pdf. Accessed June 17, 2012.

Prior to the creation of the HPP grants, preparedness and planning across healthcare facilities did not exist in most communities (Courtney et al., 2009).

Healthcare coalitions are a locally-based resilience-enhancing measure insofar as member institutions align their interests and commit their resources to conduct a cohesive, coherent medical response to the increase in, and unique needs of, patients during a public health emergency. In a major health event, individual healthcare facilities in a community need to engage effectively with one another, the larger response systems, and potentially neighboring jurisdictions. Such collaboration ensures that the personnel, supplies, and equipment distributed across otherwise autonomous facilities are applied in a systematic fashion to achieve the best medical outcomes for the community at-large (Courtney et al., 2009). Effective health care coalitions, while evolving in relation to local hazards, geography, politics, and prior institutional relationships, nonetheless exhibit an effective leadership and governance structure and strive to achieve their stated objectives (Box 5.6).

The committee saw direct evidence of the benefits of health care coalitions in discussions in Cedar Rapids, Iowa, with health care professionals affiliated with the state, county, and city. The potential for a nuclear power plant accident at a nearby facility motivated the city of Cedar Rapids and the county to establish a risk mitigation strategy for that hazard (see Box 2.4 in Chapter 2). The city's emergency planners, hospital personnel, and citizens drill four times a year along established evacuation routes in the event of a nuclear accident. These drills, including the relocation of essential medical facilities and personnel were invaluable training and were implemented during the response to the flooding of the Cedar River in the second week of June 2008.

The health care issue that has yet to be addressed is that of access to medical records of medications routinely taken and major health conditions and risks. Access is currently not readily available in emergency situations. Among the solutions discussed is a nationally linked medical record system, such as the kind already maintained by several pharmaceutical store chains, and/or a personal card containing a chip with the relevant information. Privacy issues are clearly of critical concern in these discussions, but as the post-Hurricane Katrina problems in helping patients with chronic illnesses demonstrated, the need for this information is vital.

In summary, public–private coalitions are essential for the development and execution of plans to strengthen the resilience of a community's critical infrastructure. A public–private partnership can evaluate and expand community capacity to address disaster-related risk to lifelines. Such partnerships can also help to integrate resilience into the infrastructure life cycle to ensure maintainability, sustainability, and operability of those systems before, during, and after a disaster.

BOX 5.6
New York City Preparedness Benefits from Government–Health Care Partnership

The New York City Healthcare Emergency Preparedness Program (HEPP)[a] is a coalition of hospitals, long-term-care facilities, primary care centers, emergency management services, professional associations, and medical university partners that conducts emergency preparedness activities. The coalition is coordinated with assistance from the New York City Department of Health and Mental Hygiene. The number of facilities includes 65 hospitals and acute care facilities, 400 outpatient centers, and 73 emergency medical services organizations, in addition to participants from public safety, emergency management, public health, medical societies, and hospital associations.

The goal of the program has been to create integrated and coordinated emergency planning and response in the New York Area and the coalition works toward meeting specific benchmarks such as isolation capacity, trauma care, and pharmaceutical capabilities. The program has used hazard vulnerability analysis, has developed connections to other medical facilities and city agencies, has implemented an incident command system, and has conducted training exercises and citywide drills.

[a]http://www.nyc.gov/html/doh/html/bhpp/bhpp-about.shtml.
SOURCE: Toner et al. (2009)

COMMUNICATION TO BUILD RESILIENCE

The third theme in building resilience is communication and public education, which may result in a populace that knows what hazards it faces, has the social connections that will help it endure, understands how to protect its safety and well-being, and sees itself as capable and self-sufficient. Such communications should happen at all levels, especially in promoting resilience as a national priority and a goal. However, communication and public education may be most crucial at the local level, where they strengthen social ties and capabilities, and where local knowledge and trusted relationships can amplify the power of communications. Understanding the purpose of communications is a key element in motivating resilient actions (Box 5.7).

The tactical details of risk communication—such as warning strategies, emergency communication planning, and content of messages—are vital to disaster preparedness, response, and recovery and they have been well documented elsewhere (NRC, 1989a, 2005b; Mileti and O'Brien, 1992; Mileti and Peek, 2002; Morgan et al. 2002; Fischhoff, 2009). Tactical risk communication strategies ensure timely information, reduce economic losses, prevent stigmatization, and save lives and suffering. However, communication for resilience encompasses more than tactical risk communication because resilience communication is fundamentally social, reliant upon interactions and relationships between and within communities. Communications construct how people see their roles in disasters, build the resolve necessary to endure, and encourage learning from historical precedent. Disaster planners can increase their communities' ability to plan for, absorb, and adapt to disasters by employing knowledge of specific audiences and evidence-based strategies, leveraging new media, strengthening communications networks, and helping construct disaster-resilient narratives (Table 5.3). Specific actions for this kind of communication are briefly described.

BOX 5.7
Communication That Motivates Resilient Actions

A cornerstone in communication is to know its primary objective. Is it simply to provide information without actions or is it to provide guidance on taking action? Ideally, communication should motivate individuals, families, blocks, neighborhood groups, and entire communities to develop and even rehearse plans. For example, prior to the 1994 Northridge earthquake in Los Angeles, neighborhood clusters such as blocks were encouraged to prepare as individuals and collectively. Individual preparation included earthquake kits, family communication plans, emergency lighting, etc. In Bel Air, neighbors got together and each home had a red flag and green flag. After the earthquake, people with no emergencies put a green flag in front of their homes. Designated neighbors checked houses with red flags (signaling help was needed) and without flags. The neighborhoods were essentially on their own for several days, and neighbors shared food, water, flashlights, and first-aid kits. Several houses on that street were a total loss, but there was no loss of life.

In another example, Hurricane Irene in 2011 destroyed numerous roads and bridges in upstate New York, in Vermont, in parts of Massachusetts, and in New Hampshire. In several of these states, it was difficult to determine which roads were open and which were closed. In Vermont, within 24 hours of the disaster, the Vermont Agency of Transportation had a map on the Internet with detailed information on hundreds of road closures. Essentially, it was impossible to cross from New Hampshire through Vermont to New York State for at least 30 days post-storm, but motorists and businesses could identify where they could travel and where they could not based on this kind of communication.

Source: Personal observation and experience from a committee member.

TABLE 5.3 Communication to Build Resilience: What and How

Communication Strategy	Strategy Implementation
Construct narratives that promote resilience	• Frame communities as problem solvers, individuals as capable responders • Construct narratives that reinforce social bonds, helping, and cooperation • Maintain social memory of disasters
Use evidence-based strategies for communication and public education	• Ground strategies in communities • Communicate risk • Test and evaluate efforts
Leverage social aspects of communication to strengthen ties and involve community	• Promote social interaction • Improve community use of social media networks • Improve quality, value, and trust in crowd-sourced information
Strengthen communication networks to ensure access to information	• Create multipronged, interconnected communication networks • Ensure equity in access to information

Construct Narratives that Promote Resilience

Increasing national resilience will require more than just improving communication structures and processes. To create a culture of resilience, public education and communication are important to help shift the way that Americans perceive themselves in relation to disasters and ensure that the lessons learned from our history with disasters stay active in the public's consciousness.

Communal narratives give shared experiences meaning and purpose and they demonstrate how a community sees itself and others (Alkon, 2004). By defining a group's identity and experiences and giving reason to its actions, such narratives can shape how they adapt to and recover from adversity, and thereby serve as important resources to foster resilience (Norris et al., 2008). For example, oppressed groups' positive constructions of themselves allowed them to adapt to and survive adversity (Sonn and Fisher, 1998). The extent to which communities and individuals frame themselves as capable, connected, adaptable, and self-sufficient—rather than dependent, victimized, or helpless—will affect their decision making, their actions, and their ability to cope in the face of crisis (see Box 5.3)

Top-down, command-and-control approaches to disaster management discourage community involvement, setting up expectations that only those government actors in decision-making positions can tackle the problems (NRC, 2006a). While the role of government agencies is irreplaceable, as a group of Gulf Coast community leaders and responders noted, many of the valuable responses to disaster come from the initiative and resources of individuals and communities (NRC, 2011a). A narrative shift that frames communities as the primary problem solvers and individuals as capable responders recalibrates expectations and spotlights people's innate resilient capacities (NRC, 2006a).

Norris et al. (2008) identify social linkages and a sense of community—characterized by high concern for community issues, respect for and service to others, and a sense of connection—as attributes of resilience. For example, mixed-race groups in South Africa during Apartheid maintained community resilience in the face of discrimination because of their sense of community and close bonds (Sonn and Fisher, 1998). Members of a group can strengthen their sense of community by embracing narratives that characterize the group as cohesive. Following the tragic mass shooting at Virginia Tech in 2007, Ryan and Hawdon (2008) describe how the faculty at the university accepted the administration's frame that the shooting had been an attack on the larger university community, and this in turn guided them to assume greater

BOX 5.8
Strategies to Keep Social Memory Alive

- Annual or periodic commemoration events held by community organizations, FBOs, schools, and municipalities;
- Collections of oral histories from survivors, such as the Centers for Disease Control and Prevention's "Pandemic Influenza Storybook[a]" and the "Voices After the Deluge" research by the Southern Oral History Program at the University of North Carolina at Chapel Hill;[b]
- Inclusion of local disaster histories in school curricula;
- "Digital stories" that capture people telling their personal experiences in disasters, captured on video for viewing on YouTube, Vimeo, or other websites; an illustrative example are personal stories about Hurricane Katrina captured in the Hurricane Digital Memory Bank;[c]
- Exhibits at local history and natural history museums and libraries such as museums in Cedar Rapids (Figure);
- Opportunities for intergenerational dialogue and storytelling about experiences with disasters and overcoming hardship
- Storybooks, videos, and other narrative materials that tell the stories of real disasters, such as the Survivor Tales comic books developed by the Seattle-King County Advanced Practice Center[d]

FIGURES (Left) At the African American Museum of Iowa in Cedar Rapids, the memory of the June 2008 floods that flooded the museum and damaged some of its collections is kept alive through (Center) a permanent plaque marking the high-water level inside the museum building. ☹ Right) At the Czech Museum in the Czech Village of Cedar Rapids, a timeline display documents the course of the floods over a 9-day period in June 2008. The floodwaters also reached this museum building and its collections; the high-water mark is the horizontal orange line in the top left corner of the image. Pictured are committee members and museum guide on the committee's visit to Cedar Rapids in March 2011. Source: John H. Brown, Jr., The National Academies.

[a] http://www.flu.gov/storybook/introduction'
[b] http://www.sohp.org/content/our_research/listening_for_a_change/voices_after_the_deluge/.
[c] http://hurricane.archive.org.
[d] http://www.apctoolkits.com.

responsibilities in assisting students. In this way, narratives can reinforce social bonds and also establish norms of helping, cooperation, and reciprocity. Alkon (2004) found that residents of one community internalized a narrative of themselves as people who are good at working together and were thus able to make complex policy choices despite competing interests.

Communities are only resilient insofar as they have the ability to learn from previous events and draw upon those lessons to mitigate against future events. Colten and Sumpter (2008) argue that preserving the social memory of disasters is important for resilience to take hold; they point to vital lessons about evacuation that were lost after Hurricane Betsy in 1965 that could have prevented some of the losses during Hurricane Katrina (see also NRC, 2011a; Colten and Giancarlo, 2011). When social memory is lost, communities can forget how they survived previous disasters, individuals and institutions may not retain skills needed for response and recovery, and policy makers may make decisions without regard for the hazards that exist. Maintaining social memory as a strategy for promoting resilience requires creativity by public educators and professional communicators when they draw attention to the past and its lessons for the future (Box 5.8).

Collective narratives can play a role in maintaining social memory, as they did on Simeulue Island in Indonesia, where residents orally passed down lessons learned from a devastating tsunami. When an earthquake occurred on December 26, 2004, these residents knew they had to evacuate to higher ground immediately and their island experienced far lower casualties than other neighboring islands (Meyers and Watson, 2008). In New Orleans East, the older members of the Vietnamese community transferred what they had learned from previous adversities, such as how to pool resources and how to construct homes, sharing their experiences with the younger generations. Consequently, their community recovered more quickly than other devastated parts of the region (NRC, 2011a)

Use Evidence-Based Strategies for Communication and Education

Communication strategies should be grounded in the characteristics of local communities. Audience research techniques—such as focus groups, key informant interviews, surveys, and demographic studies—will reveal what people need and want to know, leading to more effective communications than those based on assumptions (NRC 1989a). For example, in developing the California Shakeout, a large-scale public earthquake drill, planners conducted audience research that indicated that people were less interested in information about the probability of an earthquake, preferring communications that focused on what concrete actions they should take (USGS, 2008).[2]

The public is not homogenous, and no single communication approach will suffice (Bolton and Orians, 1992). Identification of personal and social characteristics of targeted audiences—such as their shared perceptions, beliefs, communication patterns, and their social contexts—will aid in the design of messages more likely to motivate behavior change (Mileti and Peek, 2002; Paton et al., 2008). To alleviate communication gaps, public educators and communicators should also examine the preexisting understandings and beliefs about disasters, hazards, and response and recovery measures held by targeted groups in comparison to experts and emergency management (Morgan et al., 2002). Understanding the differences between public and professional perspectives can identify communication gaps, especially regarding highly charged, ethical dilemmas. For example, in preparation for communicating about pandemic influenza, public engagement meetings were held in the state of Minnesota and King County, Washington, about how to ethically distribute scarce, life-saving medical resources in a crisis. By involving diverse community members and vested stakeholders, emergency planners identified

---

[2] Lucy Jones, personal communication, May 24, 2011 (see Appendix B).

similarities and differences in opinions held by each group and were able to develop targeted communication strategies (Li-Vollmer, 2010; Garrett et al. 2011).

Even with high levels of risk awareness, individuals may not translate that information to their own situation (Fitzpatrick and Mileti, 1993; Mileti and Peek, 2002). Instead, people are more likely to take protective actions if they believe those measures influence the consequences of disaster, even if they can't control the causes (Mulilis and Duval, 1995; Paton et al., 2006). In fact, whether households take protective measures depends more on how they perceive the effectiveness of those measures than on their perceptions of risk itself (Weinstein and Nicolich, 1993; Wood et al., 2011). Therefore, risk communication, a specific type of communication to build resilience, should emphasize protective actions and their benefits and also neutralize beliefs that a threat is too great for personal action to make a difference.

People are more likely to believe that their actions could make a difference when presented with messages asking them to consider helping those more vulnerable than themselves, such as children and the elderly (Paton et al., 2006). For example, Latin American immigrants, some of whom initially reported that there was no way to prepare for emergencies, said they would be motivated to develop an emergency plan for my family or to be informed so we could help others (Carter-Pokras et al., 2007). Similarly, leaders of nonprofit organizations in the Mississippi Gulf Coast advocated for messages that empower individuals to care for themselves and others, rather than those based on fear, based on their experiences helping their communities recover from Hurricane Katrina (NRC, 2011a).

People are more likely to believe that preparedness is worth the effort when they understand the potential losses that can occur from disasters, and what they can do to prevent or reduce those losses. This requires specific information about how each protective action reduces risk or contributes to safety (Paton et al., 2006; Mileti and Peek, 2002). If people are given a small number of preparedness items, starting with those easiest to adopt, they are more likely to enact them. Nonetheless, this kind of effective communication of the value of resilience represents a continual challenge for community and government leaders (see Chapter 3).

Formative testing and subsequent refining of messages and materials may help ensure that they are memorable, actionable, culturally appropriate, and comprehensible for targeted groups (Morgan et al., 2002; Andrulis et al., 2007). Using community representatives to review disaster scenarios and provide feedback on planned messaging is one approach (Paton et al., 2008). In addition, evaluation of risk communication plans following a crisis event can be used to engage the community in being part of their resilience-building strategies (NRC, 1989a).

Leverage Social Aspects of Communication to Strengthen Ties and Involve Community

When faced with uncertainty, people tend to turn to others for guidance and confirmation. Studies have found that people's interaction in their social networks can overcome passivity and have direct and indirect influence on what they know and whether they intend to take preparedness steps (Paton et al., 2008; Wood et al., 2011). To maximize communication and public education, forums encouraging community members to discuss hazard issues with ample use of visual aids, compelling media, and peer group discussion methods has been suggested (Mileti and Peek, 2002). For example, Los Angeles County Public Health and the University of California at Los Angeles have developed preparedness outreach programs using peer mentors to educate developmentally delayed adults and *promotora*[3] community health workers in the Latino community. Social media, as discussed below, offer multiple promising opportunities for promoting community planning and discussion.

The fabric and nature of community have been profoundly affected in recent years by the growth of online social media. Social networks can now grow and survive without the same ties to geography that existed in the past. Instead, electronic media allow instant communication within networks of friends (and strangers) who may be separated by long distances, and lead to a sense of community that may have little to do with geography.

So much interpersonal interaction now occurs online that the very term social network often implies a digital medium such as Facebook or Twitter. These networks can play a very important role in strengthening community by providing new ways to interact, but at the same time their lack of ties to geography may weaken local communities by diverting some of their attention elsewhere. Nevertheless, it is clear that efforts to strengthen communities and their social networks must include these new media. There is ample evidence that sites such as Craig's List can play a valuable role in helping a community's recovery by sharing information about skills and assets (Torrey et al., 2007).[4]

Individual citizens are now empowered by technology to collect and disseminate information, and such mechanisms have proven increasingly important during disasters, when reports from citizens may lead official information by minutes, and in some cases hours. Against these potential advantages the doubts about quality and the lack of the kinds of checks and confirmations of information are weighed. Goodchild and Glennon (2010), Liu and Palen (2010), Palen et al. (2010), and others have documented the role that these social media can play in collecting and sharing information about the local

---

[3] A promotora is a person who provides educational, guidance, and referral services in a community as an informal community-based worker.

[4] See also http://outreach.lib.uic.edu/www/issues/issue11_5/jones/.

situation: injuries, needs, locations of severe impact, for example. Such information is inevitably unreliable to some extent, coming as it does from volunteers who may have little training and may even have malicious intent, but it does provide immediate situational awareness. In the various wildfires that have hit the Santa Barbara area in the past few years, Goodchild and Glennon (2010) showed that volunteers can also play a vital role in synthesizing reports culled from blogs, tweets, and other postings, and reconciling apparent contradictions.

The problem of quality assurance in these situations needs specific attention. A fundamental principle of crowd sourcing argues that information is more reliable if it comes from multiple, independent sources. More effective, however, is the kind of social hierarchy used by prominent sites such as Wikipedia and Open Street Map. Individuals with a track record of reliable information are promoted through the hierarchy and play a key role in moderating and vetting reports. In essence, such systems replicate the structure of traditional government agencies, but in a manner that is consistent with their voluntary nature.

In the final analysis, however, an individual citizen experiencing the effects of a disaster must make a simple choice: to act in response to potentially unreliable but timely information provided by voluntary mechanisms, or to wait until officials are able to check and verify, by which time the impacts of the event may be severe. Efforts to strengthen communities and their use of social media, and to develop the social hierarchies that can foster trust, can do much to improve the quality, and thus increase the value, of crowd-sourced information during disasters or other traumatic events.

Strengthen Communication Networks to Ensure Access to Information

Two different mechanisms may improve communication networks to ensure access to information for resilience: (1) creation of multipronged interconnected communication networks, and (2) ensuring equity in access to information. A strong communications infrastructure can efficiently centralize collection and distribution of information and news at national, regional, and local levels before a disaster (Andrulis et al., 2007; Norris et al., 2008; Olshansky et al., 2008). This infrastructure includes the technological means to transmit information, skilled and trained human resources to carry out communication functions, and the organizational processes and social networks that facilitate the flow of communications (FEMA, 2004; NRC, 2005b; Comfort and Haase, 2006). Alternate routing and backup plans (as a part of the infrastructure planning) could prevent the type of communication breakdown that happened when Hurricane Katrina destroyed the communication system in New Orleans (Comfort and Haase, 2006). Plans for communication that maintain parity with the technologies that the public widely uses, such as text messaging and social media, are also important (Karasz and Bogan, 2011; Merchant et al., 2011), as are nonelectronic forms of communication such as door-to-door provision of information, distribution of brochures, and meetings at community centers in the event of power failures and for those who lack easy access to online communication.

Flexibility in the face of the unknown is vital to a communication network that can adapt to changing circumstances. Reliance on rigid command-and-control strategies for communication can prove detrimental; instead, building multipronged networks that feed into and pull from many community nodes may constitute a better communication strategy (Norris et al., 2008). Dense communication networks contribute to community action because individuals tend to confirm information across multiple sources and within their social spheres before determining courses of action (Wood et al., 2011). Inclusion of CBOs—along with local, state, and federal agencies and response partners—creates more avenues for rapidly delivering critical information. More importantly, incorporating CBOs also leverages sources of information that are already trusted in their communities, resulting in better outreach to diverse populations and more effective coordination of communications (Andrulis et al., 2007). A more inclusive communications network also creates a feedback loop that circulates communities' needs from the communities to leaders and helps set realistic expectations from leaders to communities (Schoch-Spana et al., 2007). An authentic two-way flow of communications builds trust in public information campaigns and the public's willingness to take needed actions (NRC, 1989a; Paton et al., 2008).

A second component of the communications network is recognizing and addressing inequities in access to information that result from culture, language, socioeconomic status, functional ability, literacy, and trust (Kasperson et al., 1992; Vaughn, 1995; Andrulis et al., 2007). When these communication barriers are not addressed, equal access to food, medical treatment, safety information, and other lifesaving resources cannot be assured (Fothergill et al., 1999; Carter-Pokras et al., 2007). For example, failure to provide evacuation orders in multiple languages, culturally competent ways, or through adequately targeted channels has led to endangerment and unnecessary deaths among ethnic minority and immigrant groups (Muñiz, 2006; Spence et al., 2007). People who have difficulty accessing needed care and resources day to day are at even higher risk from disasters, and failure to ensure equity in access to information can further amplify the hardships these individuals face.

Communication networks that include diverse stakeholders are fundamental to reaching more diverse populations. People working in specific communities often have the expertise and relationships in place to best communicate to the families and individuals they serve. When trusted sources from the community act as messengers, the information is more likely to be received, understood, and accepted than if it comes from an unknown or government source (Fothergill, 1999; Mileti and Peek, 2002; Muñiz, 2006;

Andrulis et al., 2007;). Trusted community sources include ethnic media, FBOs, health care providers, community leaders, and CBOs (Andrulis et al., 2007; Carter-Pokras, 2007). The Aware & Prepare Initiative in Santa Barbara County[5] is an example of a public–private partnership to enable nonprofit organizations and government agencies to work together on disaster resilience-building measures. A particular focus of the public education and awareness segment of this initiative is on communicating directly with vulnerable populations (J. Moreno, personal communication, May 24, 2011; Appendix B).

As standard protocol, communications and educational materials must be available in multiple languages and in translation (Mileti and Peek, 2002). Translation alone may be insufficient, and the review by individuals from target communities to ensure cultural adoption and the ability of the materials to meet needs of people with lower literacy or different functional abilities can ensure that the messaging is appropriate and acceptable and is absorbed and adopted by the intended audiences (Mileti and Peek, 2002; Andrulis et al., 2007).

ZONING AND BUILDING CODES AND STANDARDS

Local communities have a variety of mechanisms at their disposal to reduce risks and enhance resilience—mechanisms that are largely under the control of local jurisdictions. Among the most basic of these are land use, zoning, and building codes and standards (see also Chapter 2 under "Nonstructural Measures" as part of risk management planning and implementation).

Zoning and Building Codes and Standards to Strengthen Community Resilience

Building codes set the minimum requirements for infrastructure and are established through a hierarchy of national, regional, and local governments. Codes and standards exist to guide construction of residential, commercial, and industrial buildings, and to inform zoning and land-use considerations (Ching and Winkel, 2009). Building codes can support resilience by helping to prevent or minimize damage to the built environment during natural disasters; minimum standards of siting and construction can also help ensure public health and safety. However, a balance between adding to the codes to protect infrastructures from disasters and causing the cost of buildings to increase to a point where the costs prevent or delay new construction are considerations that decision makers, the private sector, and community have to take into account. Also, if adjacent communities adopt or enforce building codes differently,

---

[5] See also http://www.orfaleafoundation.org/partnering-impact/collective-impact-initiatives/aware-prepare.

developers may choose to develop in the community with lower requirements in order to save money on construction. Such discrepancies may call for increased regional or statewide consistency in the use of building codes. Additionally, the federal government constructs its buildings to meet a set of federal codes, and maintaining a balance between federal and local codes and standards is also challenging. For example, an NRC report found that "designs for federal buildings were inappropriate to local conditions and resulted in costly difficulties during construction that could have been avoided had local building code provisions been updated to reflect the model codes" (NRC, 1989b, p. 10). Presently, high-level resilience is not addressed in these minimum requirements for the codes, resulting in limited design guidance available to the community on providing enhanced safety to the built environment (NIBS and DHS, 2010).

Background on and Purpose of Codes and Standards in Resilience

National codes provide a base upon which regional and, subsequently, local codes are developed. This base lays the groundwork for a minimum level of resilience to be set at a national level, with room for specific updates at the regional and local scales. The origin of the building codes used today lies in the fires that damaged American cities throughout the 1800s and were initially written to support the needs of insurance companies for fire protection and hazard reduction (NIBS and DHS, 2010).[6] This fire-based foundation of building codes can be considered an initial step toward establishing resilience. The codes are written in such detail that specifications for means of exiting from a building are included (Ching and Winkel, 2009). At the core, the codes are designed to protect health and life—providing safe passage for individuals if a building should collapse. The minimum standards for codes do not consider the structure's performance or hazard resilience in a specific way, although stricter codes may be developed to consider these aspects of a structure (Box 5.9).

Most communities adhere to the International Code Council's (ICC) International Codes (or I-codes), which provide minimum standards for building and fire safety.[7] Codes also provide a consistent set of standards for residential and commercial buildings across the nation. Model codes published by ICC are adopted, sometimes in modified form, by the legislatures of individual U.S. states and carry the force of law. These codes include[8]:

- International Building Code,
- International Residential Code for One- and Two-Family Dwellings,

---

[6] See http://www.iccsafe.org/CS/Pages/default.aspx. Accessed February 11, 2012.

[7] See http://www.iccsafe.org/AboutICC/Pages/default.aspx for more information on the International Code Council's history and its guidelines. Accessed February 8, 2012.

[8] International Code Council, http://www.iccsafe.org/AboutICC/Pages/default.aspx.

- International Existing Building Code,
- International Fire Code,
- International Zoning Code, and
- International Wildland-Urban Interface Code.

Building code enforcement, however, is generally the responsibility of local government, which hires building inspectors to ensure their implementation. Building codes have been shown repeatedly to be effective in reducing property damage, preserving human life, and increasing the resilience of communities (Multihazard Mitigation Council, 2005; see also Box 5.9). However, except where federally owned property or interests are involved, the federal government has little role in establishing local building codes and standards, or zoning laws (see below). Thus, the adoption and enforcement of building codes and standards lie predominantly at the local level, and are highly variable across the nation. Rigorous enforcement of updated building codes continues to be one of the surest mechanisms for improving resilience of infrastructure.

BOX 5.9
Wind Resistance Building Codes
Helped Floridians Weather Hurricane Charley

The devastation wrought by Hurricane Andrew when it struck Florida in 1992 triggered a reevaluation of existing building code standards and their enforcement. In 1995, coastal areas of Florida started to use and enforce high-wind design provisions for residential housing, including those that ensured that all loads were directed to the foundations. Builders and building officials received extensive training in concert with this development. In the late 1990s, the state of Florida moved toward adopting a statewide building code, something that was achieved in 2002. This was accompanied by the training of all licensed engineers, architects, and contractors in the new code. In 2004, four major hurricanes, the first of which was Hurricane Charley, pummeled Florida from both coastlines over a period of 6 weeks. A study of losses in the hardest hit area, Charlotte County (which had implemented high-wind standards in 1996) revealed that enforcement of modern engineering design-based building codes significantly enhanced the performance of residential homes during Hurricane Charley. Charlotte County policyholders for homes built after 1996 filed 60 percent fewer claims than those for homes built before 1996; when a loss did occur for a post-1996 home, the claim was 42 percent less severe than that for a pre-1996 home. The study also concluded that the new building code requirements permitted homeowners to return to their residences more quickly, thus reducing the disruption to their daily lives.

SOURCE: Institute for Business and Home Safety's Building Code Resources (2004).

In a similar manner, zoning laws reduce the vulnerability and impacts of disaster in a community by preventing the development of communities in places exposed to hazards. Zoning laws are the responsibility of local, regional, or state authorities, depending upon the specific setting and agreements among authorities. The authority for zoning laws generally lies with the city or county government, though agreements among jurisdictions may assign authority to a metropolitan or regional commission.

The first municipality in the United States to develop a zoning law was New York City, which implemented its groundbreaking Zoning Resolution of 1916 in response to competing public needs related to urban development (New York City, 2011). Though zoning laws developed slowly over the following 100 years, and some provisions of zoning laws are contentious and have been tested and challenged in the courts, it is widely recognized that thoughtful land-use planning combined with zoning laws constitute a very effective set of tools for keeping citizens and their property, to some extent, out of harm's way (Burby, 1998; see also Chapter 2).

A recent example of such a law is the new zoning code adopted by New Orleans in 2011, six years after the events associated with Hurricane Katrina (Box 5.10). The new master plan for development in New Orleans even contains a chapter dedicated to community resilience and has, as one of its goals, a broad and encompassing community standard of resilience with respect to flooding and other hazards.[9] This zoning code also explicitly recognizes the valuable role of natural defenses to natural disasters. Clearly, effective community land-use planning and zoning are fundamental to building resilience.

BOX 5.10
New Orleans' New Zoning Code

According to New Orleans' new ordinance, the purpose of zoning is

1. To encourage and promote, in accordance with present and future needs, the safety, morals, health, order, convenience, prosperity, and general welfare of the citizens of the City of New Orleans;
2. To provide for efficiency and economy in the process of development;
3. To provide for the appropriate and best use of land;
4. To provide for preservation, protection, development, and conservation of the natural resources of land, water, and air;
5. To provide for adequate public utilities and facilities, and for the convenience of traffic and circulation of people and goods;

[9] See Chapter 12 of the "Plan for the 21st Century: New Orleans 2030," available at http://www.nolamasterplan.org/documentsandrresources.asp#C3.

6. To provide for the safe use and occupancy of buildings and for healthful and convenient distribution of population;
7. To provide for promotion of the civic amenities of beauty and visual interest, for preservation and enhancement of historic buildings and places, and for promotion of large-scale developments as means of achieving unified civic design; and
8. To provide for development in accord with the Comprehensive Plan.

SOURCE: New Orleans Comprehensive Zoning Ordinance, March 3, 2011, http://library.municode.com/index.aspx?clientId=16306&stateId=18&stateName=Louisiana.

Consequences of a Lack of Building Code Enforcement and Zoning Provisions

Despite widespread availability of codes and zoning guidelines and agreement by most officials that these governance tools benefit community resilience, many unsafe buildings still exist and many communities continue to allow development in hazardous areas. The major reasons that municipal and state jurisdictions find it difficult to enforce building codes and zoning laws include the lack of resources or number of qualified personnel to do so, pressure from developers to grow communities, and lack of political will to manage land use through zoning (Burby, 1998).

Building code enforcement costs money, namely in the form of salaries for qualified, trained technical staff who inspect both new and retrofit construction, issue judgments on compliance, and carry out follow-up inspections when failure to comply arises. Municipal and county governments facing limited budgets, and many competing public demands often result in cuts to these critical personnel. As expressed by useful-community-development.org, "Most towns and cities practice only complaint-based code enforcement, largely for cost reasons."[10] Construction and building inspectors held about 106,400 jobs in 2008, and the median annual wages of construction and building inspectors were $50,180 in May 2008 (U.S. Bureau of Labor Statistics, 2009). Many of the 19,510 incorporated towns and cities in the United States (U.S. Census Bureau, 2010) struggle to maintain the most basic public services delivered by police, fire, and teachers. At the same time that inspectors are in short supply, the builders and building owners may resist compliance, especially if such measures require additional investment. Though the short-term funding issues are unfortunately often the determinant of local code enforcement, the adoption and enforcement of building codes have proven to be economically beneficial in reducing property damage, improving life safety, and increasing the resilience of communities (Cohen and Noll, 1981; Multihazard Mitigation

---

[10] http://www.useful-community-development.org/code-enforcement.html.

Council, 2002). However, tension between local and national interests arises when local building codes contain provisions that respond to specific community interests and concerns. The national code may be seen as a constraint on the community's ability to construct buildings the way that they require (NRC, 1989b).

Strategies to Reverse Lack of Enforcement

Existing engineering technologies, tools, and design criteria provide guidance for codes and standards to support prevention, mitigation, and risk avoidance; however, accelerating the enforcement of these regulations has proved to be difficult and expensive for local government. What is the best way to encourage and accelerate the enforcement of building and zoning codes where enforcement is currently not universal? One potential mechanism is to tie the adoption and enforcement of building codes to state eligibility requirements for federal disaster relief funds and programs. Although sometimes politically unpopular, such an approach can help build a culture of resilience. Other mechanisms may include the provision of additional training to public safety officials for code enforcement inspections(e.g., fire departments, emergency services personnel, emergency managers) who could assist in tight fiscal times (Timm, 2004). Finally, penalties and sanctions levied against developers who blatantly ignore codes is another option, but this may also result in the need for more inspections and the resources to hire additional staff.

To address resilience in the built environment, codes and standards may also need to consider integrating new language, considering all of the building design criteria, and expanding standards beyond life-safety aspects, including safety and usability (Poland, 2011). Performance-based standards and codes, for example, have historically served as objective-based requirements for a building designer to meet (Ching and Winkel, 2009). New building codes and standards that extend beyond life-safety aspects may include resilient design concepts in a performance-based approach, as well as continuity of operations (NIBS and DHS, 2010). Additionally, the codes could integrate frequent and well-adopted design measurements and standards, providing a flexible platform to address different facility and structure types and recognizing the differing levels of performance that are required.

Higher minimums for building codes may be another mechanism to increase the visible, direct links between building code and standard enforcement and resilience. The current minimum requirements prescribed by building codes, while laying the groundwork for resilience, do not provide adequate design guidance for resilience. An outcome of the *Designing for a Resilient America: A Stakeholder Summit on High Performance Resilient Buildings and Related Infrastructure* held in November 2010 was that U.S.

building codes and standards need to set more stringent minimum requirements, for health and life safety, that are enforced by many jurisdictions across the country and supported by state legislation.[11] Design guidance on providing serviceability criteria and enhanced safety standards is limited or, in some cases, unavailable to designers and owners because higher resiliency requirements are not integrated at the most minimum model building codes. Uniform adoption by jurisdictions begins with the development of design criteria, building codes, and standards that address resiliency objectives and the technologies and validation for their use (NIBS and DHS, 2010).

RESEARCH AND INFORMATION NEEDS

A number of areas need additional research to fully understand local opportunities for and constraints to enhancing community resilience. First, no systematic or evidence-based assessment has been conducted to identify which strategies are most effective in fostering local collaborations to build community resilience. Most of the information appears to be anecdotal or tied to case studies at present, with little evidence to support whether generic strategies can be customized for the local context. Second, the economic impacts of changes in building codes or zoning laws are not tied well or directly to the receipt of disaster relief. Would such explicit ties make communities more receptive to implementation and/or enforcement of building codes and zoning laws? At present, that question cannot be answered. Finally, studies are needed to evaluate the reliability and validity of information communicated through social media and whether the integration of social media into disaster preparedness, response, and resilience efforts affects the costs, quality, or outcomes (Merchant et al., 2011).

SUMMARY AND RECOMMENDATIONS

Resilience requires reinforcement of our physical environment—the buildings and critical infrastructure that support the communities in which we live. It also requires the strengthening of our social infrastructure—the local community networks that can mobilize to plan, make decisions, and communicate effectively. The interconnectedness of the social and physical infrastructure requires that both aree enhanced simultaneously with equal consideration to increasing resilience. The principal action through which a local community could vastly accelerate progress toward enhanced resilience of its

---

[11] For more information on the U.S. Department of Homeland Security stakeholder summit, please see http://www.dhs.gov/files/publications/st-bips-designing-resilient.shtm. Accessed February 12, 2012.

social and physical infrastructure is establishment of a problem-solving coalition of local leaders from public and private sectors, with ties to and support from federal and state governments, and with input from the greater citizenry. The charge of such a coalition would be to assess the community's exposure and vulnerability to risk, educating and communicating about risk, and evaluating and expanding its capacity to handle such risk. A truly robust coalition would have at its core a strong leadership and governance structure, with a person or persons with adequate time, skill, and dedication necessary for the development and maintenance of relationships among all partners.

Recommendation: **Federal, state, and local governments should support the creation and maintenance of broad-based community resilience coalitions at local and regional levels.** Efforts to support coalition development should include:

- Assessment by the Department of Homeland Security and Department of Health and Human Services—to the extent that these two agencies administer state and local grant programs to bolster national preparedness capabilities—of present federal funding frameworks and technical guidance. Such an assessment could gauge whether communities have sufficient support and incentive to adopt collaborative problem-solving approaches toward disaster resilience and disaster risk management.
- Adoption by communities of collaborative problem-solving approaches in which all private and public stakeholders (e.g., businesses, NGOs, CBOs, and FBOs) are partners in identifying hazards, developing mitigation strategies, communicating risk, contributing to disaster response, and setting recovery priorities. The emergency management community is an important integrated part of these discussions, potentially taking on a leadership role.
- Commitment by state and local governments to secure adequate personnel to create and sustain public–private resilience partnerships, to promulgate and implement proposed national resilience standards and guidelines for communities, and to assist communities in the completion of the proposed national resilience scorecard.

Building codes and standards are effective in mitigating and reducing disaster risk to communities. However, codes and standards have some variability due to the nature of local hazards; across the nation they are unevenly enforced and many people do not know they exist. In addition to codes and standards, guidelines, certifications, and practices also can be effective in fostering resilience.

Recommendation: **Federal agencies, together with local and regional partners, researchers, professional groups, and the private sector should**

develop an essential framework (codes, standards, and guidelines) that drive the critical structural functions of resilience.

This framework should include national standards for infrastructure resilience and guidelines for land use and other structural mitigation options, especially in known hazard areas such as floodplains. The Department of Homeland Security is an appropriate agency to help coordinate this government-wide activity. The adoption and enforcement of this framework at the local level should be strongly encouraged by the framework document and accompanied by a commitment from state and local governments to ensure that zoning laws and building codes are adopted and enforced.

REFERENCES

ALA (American Lifelines Alliance). 2005. Protecting Our Critical Infrastructure: Findings and Recommendations from the American Lifelines Alliance Roundtable. Washington, DC: National Institute of Building Sciences. Available at http://www.americanlifelinesalliance.com/pdf/RoundtableReportWebPosted081705.pdf.

Alkon, A. H. 2004. Place, stories, and consequences: Heritage narratives and the control of erosion on Lake County, California vineyards. *Organization and Environment* 17(2):145-169.

Andrulis, D. P., N. J. Siddiqui, and J. L. Gantner. 2007. Preparing racially and ethnically diverse communities for public health emergencies. Health Affairs (Millwood) 26(5):1269-1279.

Avery, G. H., and J. Zabriskie-Timmerman. 2009. The impact of federal bioterrorism funding programs on local health department preparedness activities. *Evaluation and the Health Professions* 32(2):95-127.

Barkley, C. 2009. Lifelines: Upgrading Infrastructure to Enhance San Francisco's Earthquake Resilience. San Francisco, CA: San Francisco Planning and Urban Research Association. Available at http://www.spur.org/files/lifelines-formatted.11.10.pdf.

Bolton, P. A., and C. E. Orians. 1992. Earthquake Mitigation in the Bay Area: Lessons from the Loma Prieta Earthquake. Seattle, WA: Battelle Human Affairs Research Centers.

Broom, J. 2007. Carbon monoxide: Last year's surprise killer still claims lives. *The Seattle Times,* December 13. Available at http://seattletimes.nwsource.com/html/localnews/2004068680_carbon13m.html.

Bruneau, M., S. E. Chang, R. T. Eguchi, G. C. Lee, T. D. O'Rourke, A. M. Reinhorn, M. Shinozuka, K. Tierney, W. A. Wallace, and D. von Winterfeldt. 2003. A framework to quantitatively assess and enhance the seismic resilience of communities. *Earthquake Spectra* 19(4):733-752.

Burby, R. J., ed. 1998. *Cooperating with Nature: Confronting Natural Hazards with Land-Use Planning for Sustainable Communities.* Washington, DC: Joseph Henry Press.

Carter-Pokras, O., R. E. Zambrana, S. E. Mora, and K. A. Aaby. 2007. Emergency preparedness: Knowledge and perceptions of Latin American immigrants. *Journal of Health Care for the Poor and Underserved* 18(2):465-481.

CDC (Centers for Disease Control and Prevention). 2011a. A National Strategic Plan for Public Health Preparedness and Response. Available at http://www.cdc.gov/phpr/publications/2011/A_Natl_Strategic_Plan_for_Preparedness_20110901A.pdf. Accessed June 17, 2012.

CDC . 2011b. Public Health Preparedness: 2011 State-by-State Update on Laboratory Capabilities and Response Readiness Planning. Available at http://www.cdc.gov/phpr/pubs-links/2011/documents/SEPT_UPDATE_REPORT_9-13-2011-Final.pdf. Accessed June 17, 2012.

CDC. 2011c. Public Health Preparedness Capabilities: National Standards for State and Local Planning. Available at http://www.cdc.gov/phpr/capabilities/DSLR_capabilities_July.pdf. Accessed June 17, 2012.

CDC. 2012. People at High Risk of Developing Flu-Related Complications. Available at http://www.cdc.gov/h1n1flu/highrisk.htm. Accessed June 16, 2012.

Chakos, A., P. Schultz, and L. T. Tobin. 2002. Making it work in Berkeley: Investing in community sustainability. *Natural Hazards Review* 3(2):55-67.

Chang, S. E. 2009. Infrastructure resilience to disasters. *The Bridge* 39(4):36–41. Available at www.nae.edu/File.aspx?id=17673.

Ching F. D. K., and S. R. Winkel. 2009. *Building Codes Illustrated: A Guide to Understanding the International Building Code.* Hoboken, NJ: John Wiley & Sons.

Cohen, L., and R. Noll. 1981. The economics of building codes to resist seismic shock. *Public Policy* 29(1):1-29.

Colten, C. E., and A. Giancarlo. 2011. Losing resilience on the Gulf Coast: Hurricanes and social memory, *Environment: Science and Policy for Sustainable Development* 53(4):6-18.

Colten, C. E., and A. R. Sumpter. 2008. Social memory and resilience in New Orleans. *Natural Hazards* 48(3):355-364.

Comfort, L. K., and T. W. Haase. 2006. Communication, coherence, and collective action: The impact of Hurricane Katrina on communications infrastructure. *Public Works Management and Policy* 10 (4):328-343.

Courtney, B., E. Toner, R. Waldhorn, C. Franco, K. Rambhia, A. Norwood, T. V. Inglesby, and T. O'Toole. 2009. Healthcare coalitions: The new foundation for national healthcare preparedness and response for catastrophic health emergencies. *Bioterrorism and Biosecurity* 7(2):153-163.

DHHS (Department of Health and Human Services). 2009. National Health Security Strategy. Available at http://www.phe.gov/Preparedness/planning/authority/nhss/strategy/Documents/nhss-final.pdf.

DHS (Department of Homeland Security). 2010. Quadrennial Homeland Security Review Report: A Strategic Framework for a Secure Homeland. Available at: http://www.dhs.gov/xlibrary/assets/qhsr_report.pdf.

Enarson, E. 2007. Identifying and addressing social vulnerabilities. In *Emergency Management: Principles and Practice for Local Government,* 2nd Ed. Washington, DC: ICMA Press, pp. 257-278.

FEMA (Federal Emergency Management Agency). 2004. National Response Plan. Available at http://www.fema.gov/emergency/nrf/

FEMA . 2010. Comprehensive Preparedness Guide 101: Developing and Maintaining Emergency Operations Plans, Version 2.0. Available at http://www.fema.gov/pdf/about/divisions/npd/CPG_101_V2.pdf.

FEMA. 2011. A Whole Community Approach to Emergency Management: Principles, Themes, and Pathways to Action. FDOC 104-008-1. Available at http://www.fema.gov/library/viewRecord.do?id=4941.

Fischhoff, B. 2009. Risk perception and communication. In *Oxford Textbook of Public Health,* 5th Ed.. Oxford: Oxford University Press, pp. 940-952.

Fitzpatrick, C., and D. S. Mileti. 1994. Public risk communication. In *Disasters, Collective Behavior, and Social Organization*, R. R. Dynes and K. J. Tierney, eds. Newark: University of Delaware Press, pp. 71-84.

Fothergill, A., E. G. M. Maetas, and J. D. Darlington. 1999. Race, ethnicity and disasters in the United States: A review of the literature. *Disasters* 23(2):156-173.

Franco, C., and T. K. Sell. 2012. Federal agency biodefense funding, FY2012-FY2013. *Biosecurity and Bioterrorism* 10(2):162-181.

Flynn, S. E. 2008. America the Resilient: Defying Terrorism and Mitigating Natural Disasters. Foreign Affairs. Tampa, FL: Council on Foreign Relations. Available at www.foreignaffairs.com/articles/63214/stephen-e-flynn/america-the-resilient.

Garrett, J. E., D. E. Vawter, K. G. Gervaise, A. W. Prehn, D. A. DeBruin, F. Livingston, A. M. Morley, J. Liaschenko, and R. Lynfield. 2011. The Minnesota Pandemic Ethics Project: Sequenced, robust public engagement processes. *Journal of Participatory Medicine* 3. Available at http://www.jopm.org/evidence/research/2011/01/19/the-minnesota-pandemic-ethics-project-sequenced-robust-public-engagement-processes/.

Geis, D. E. 2000. By design: the disaster resistant and quality of-of-life community. *Natural Hazards Review* 1(3):151-160.

Goodchild, M. F., and J. A. Glennon. 2010. Crowdsourcing geographic information for disaster response: A research frontier. *International Journal of Digital Earth* 3(3):231-241.

Granger, M. M., B. Fischoff, A. Bostrom, and C. J. Atman. 2002. *Risk Communication: A Mental Models Approach.* New York: Cambridge University Press.

Guikema, S. 2009. Infrastructure design issues in disaster-prone regions. *Science* 323(5919):1302-1303.

HRSA (Health Resources and Services Administration). 2002. Bioterrorism Hospital Preparedness Program, Cooperative Agreement Guidance. Available at http://www.aha.org/content/00-10/BioPrepCoopAgGuid.pdf.

Honore, R. L. 2008. Health disparities: Barriers to a culture of preparedness. *Journal of Public Health Management Practice* 14(Suppl.):S5-S7.

IAEM (International Association of Emergency Managers). 2007. Principles of Emergency Management Monograph (Supplement). Available at https://www.iaem.com/EMPrinciples/index.htm.

Institute for Business and Home Safety's Building Code Resources. 2004. The Benefits of Modern Wind Resistant Building Codes on Hurricane Claim Frequency and Severity—A Summary Report: Hurricane Charley, Charlotte County, Florida, August 13. Available at http://www.disastersafety.org/content/data/file/ibhs_building_code_kit.pdf.

IOM (Institute of Medicine). 2008. *Research Priorities in Emergency Preparedness and Response for Public Health Systems: A Letter Report*. Washington, DC: The National Academies Press.

Karasz, H., and S. Bogan. 2011. What 2 know b4 u text: Short Message Service options for local health departments. *Washington State Journal of Public Health Practice* 4(1):20. Availaboe at http://www.wsphajournal.org/V4N1/V4N1_Karasz.pdf.

Kasperson, R. E., D. Golding, and S. Tuler. 1992. Social distrust as a factor in sitting hazardous facilities and communication risks. *Journal of Social Issues* 48:161-187.

Kessler, R. C. 2007. Hurricane Katrina's impact on the care of survivors with chronic medical conditions. *Journal of General Internal Medicine* 22:1225-1230

Li-Vollmer, M. 2010. Health care decisions in disasters: Engaging the public on medical service prioritization during a severe influenza pandemic. *Journal of Participatory Medicine* 2. Available at http://www.jopm.org/evidence/casestudies/2010/12/14/health-care-decisions-in-disasters-engaging-the-public-on-medical-service-prioritization-during-a-severe-influenza-pandemic/.

Liu, S. B. and L. Palen. 2010. The new cartographers: Crisis map mashups and the emergency of nongeographic practice. *Cartography and Geographic Information Science* 37(1):69-90.

Lurie, N. 2009. H1N1 influenza, public health preparedness, and health care reform. *New England Journal of Medicine* 361(9):843-845.

Meo, M., B. Ziebro, and A. Patton. 2004. Tulsa turnaround: From disaster to sustainability. *Natural Hazards Review* 5(1):1-9.

Merchant, R. M., S. Elmer, and N. Lurie. 2011. Integrating social media into emergency-preparedness efforts. *New England Journal of Medicine* 365:289-291.

Meyers, K., and P. Watson. 2008. Legend, ritual, and architecture on the Ring of Fire. In *Indigenous Knowledge for Disaster Risk Reduction: Good Practices and Lessons Learned from Experiences in the Asia-Pacific Region*. Bangkok: United Nations International Strategy for Disaster Reduction, pp. 17-22.

Mileti, D. S., and P. W. O'Brien. 1992. Warnings during disaster: Normalizing communicated risk. *Social Problems* 39(1):40-57.

Mileti, D. S., and L. A. Peek. 2002. Understanding individual and social characteristics in the promotion of household disaster preparedness. In *New Tools for Environmental Protection: Education, Information, and Voluntary Measures*. Washington, DC: The National Academies Press, pp. 125-140.

Morrow, B. H. 2008. Community Resilience: A Social Justice Perspective. CARRI Research Report No. 4. Available at http://www.resilientus.org/library/FINAL_MORROW_9-25-08_1223482348.pdf.

Mulilis, J.-P., and T. S. Duval. 1995. Negative threat appeals and earthquake preparedness: A person-relative-to-event (PrE) model of coping with threat. *Journal of Applied Social Psychology* 25(15):1319-1339.

Multihazard Mitigation Council. 2002. Parameters for an Independent Study to Assess the Future Benefits of Hazard Mitigation Activities, Prepared for the Federal Emergency Management Agency by the Panel on Assessment of Savings from Mitigation Activities, Multihazard Mitigation Council. Washington, DC: National Institute of Building Sciences.

Multihazard Mitigation Council. 2005. Natural Hazard Mitigation Saves: An Independent Study to Assess the Future Savings from Mitigation Activities, Vol. 2: Study Documentation, p. 43. Available at http://nibs.org/MMC/mmchome.html.

Muñiz, B. 2006. *In the Eye of the Storm: How the Government and Private Response to Hurricane Katrina Failed Latinos*. Washington, DC: National Council of La Raza.

National Infrastructure Advisory Council. 2009a. Critical Infrastructure Resilience: Final Report and Recommendations. Appendix D contains an extensive bibliography on resilience. Available at www.dhs.gov/xlibrary/assets/niac/niac_critical_infrastructure_resilience.pdf.

National Infrastructure Advisory Council. 2009b. Framework for Dealing with Disasters and Related Interdependencies: Final Report and Recommendations. Available at www.dhs.gov/xlibrary/assets/niac/niac_framework_dealing_with_disasters.pdf.

National Infrastructure Advisory Council. 2010. A Framework for Establishing Critical Infrastructure Resilience Goals: Final Report and Recommendations by the Council. Available at http://www.dhs.gov/xlibrary/assets/niac/niac-a-framework-for-establishing-critical-infrastructure-resilience-goals-2010-10-19.pdf.

NIBS and DHS (National Institute of Building Sciences and Department of Homeland Security). 2010. *Designing for a Resilient America: A Stakeholder Summit on High Performance Resilient Buildings and Related Infrastructure*. Washington, DC: NIBS.

NRC (National Research Council). 1989a. *Improving Risk Communication*. Washington, DC: The National Academies Press.

NRC. 1989b. *Use of Building Codes in Federal Agency Construction*. Washington, DC: The National Academies Press.

NRC. 2005a. *Creating a Disaster Resilient America: Grand Challenges in Science and Technology—Summary of a Workshop*. Washington, DC: The National Academies Press.

NRC. 2005b. *Public Health Risks of Disasters: Communication, Infrastructure, and Preparedness—Workshop Summary*, Roundtable on Environmental Health Sciences, Research, and Medicine, W. Hooke and P. G. Rogers, eds. Washington, DC: The National Academies Press.

NRC. 2006a. *Community Disaster Resilience: A Summary of the March 20, 2006 Workshop of Disasters Roundtable*. Washington, DC: The National Academies Press.

NRC. 2006b. *Facing Hazards and Disasters: Understanding Human Dimensions*. Washington, DC: The National Academies Press.

NRC. 2009. *Applications of Social Network Analysis for Building Community Disaster Resilience: Workshop Summary*. Washington, DC: The National Academies Press.

NRC. 2010. *Private-Public Sector Collaboration to Enhance Community Disaster Resilience: A Workshop Report*. Washington DC: The National Academies Press.

NRC. 2011a. *Assessing National Resilience to Hazards and Disasters: The Perspective from the Gulf Coast of Louisiana and Mississippi: Summary of a Workshop*. Washington, DC: The National Academies Press.

NRC, 2011b. *Building Community Disaster Resilience Through Private-Public Collaboration*. Washington, DC: The National Academies Press.

New York City Department of City Planning. 2011. About Zoning. Available at: http://www.nyc.gov/html/dcp/html/zone/zonehis.shtml.

Norris, F. H., S. P. Stevens, B. Pfefferbaum, K. F. Wyche, and R. L. Pfefferbaum. 2008. Community resilience as a metaphor, theory, set of capacities, and strategy for disaster readiness. *American Journal of Community Psychology* 41:127-150.

Nuzzo, J. 2009. Preserving gains from public health emergency preparedness cooperative agreements. *Biosecurity and Bioterrorism* 7(1):35-36.

Orians, C., S. Rose, B. Hubbard, J. Sarisky, L. Reason, T. Bernichon, E. Liebow, B. Skarpness, and S. Buchanan. 2009. Strengthening the capacity of local health agencies through community-based assessment and planning. *Public Health Reports* 124:875-882.

Olshansky, R. B., L. A. Johnson, J. Horne, and B. Nee. 2008. Planning for the rebuilding of New Orleans. *Journal of the American Planning Association* 74(3):273-287.

Palen, L., K. M. Anderson, G. Mark, J. Martin, D. Sicker, M. Palmer, and D. Grunwald, 2010. A vision for technology-mediated support for public participation and assistance in mass emergencies and disasters. In *Proceedings of the Association of Computing Machinery and British Computing Society's 2010 Conference on Visions of Computer Science*. Available at http://www.cs.colorado.edu/~palen/computingvisionspaper.pdf.

Paton, D., J. McClure, and P. T. Burgelt. 2006. Natural hazard resilience: The role of the individual and household preparedness. In *Disaster Resilience: An Integrated Approach,* D. Paton and D. Johnston, eds. Springfield, IL: Charles C Thomas, pp. 105-127.

Paton, D., B. Parkes, M. Daly, and L. Smith. 2008. Fighting the flu: Developing sustained community and resilience. *Health Promotion Practice* 9(4):45S-53S.

Patton, A. 2007. Collaborative emergency management. In *Emergency Management: Principles and Practice for Local Government*, 2nd ed. Washington, DC: ICMA Press, pp. 71-85.

Pfefferbaum, R. L., D. B. Reissman, B. Pfefferbaum, K. F. Wyche, F. H. Norris, and R. W. Klomp. 2008. Factors in the development of community resilience to disasters. In *Intervention and Resilience After Mass Trauma*, M. Blumenfeld, ed. Cambridge, UK: Cambridge University Press, pp. 49-68.

Poland, C. 2009. The Resilient City: Defining What San Francisco Needs from Its Seismic Mitigation Policies. San Francisco, CA: San Francisco Planning and Urban Research Association. Available at: http://www.spur.org/files/u7/RC-overarching.pdf.

Poland, C. 2011. Community planning for resilience SPUR: Standards for disaster resilience for buildings and physical infrastructure systems. Presentation at the ANSI-HSSP 2011 Workshop, Arlington, VA.

Port of Los Angeles. 2011. Annual Sustainability Report. http://www.portoflosangeles.org/DOC/REPORT_Port_Sustainability_Report_2011.pdf

Prater, C. S., and M. K. Lindell. 2000. Politics of hazard mitigation. *Natural Hazards Review* 1(2):73-82.

Public Health–Seattle and King County. 2006. Windstorm 2006: After Action Report, December 14-22. Available at http://www.apctoolkits.com/vulnerablepopulation/case/detail/communications_with_immigrant_and_lep_communities_in_king_county_washington/.

Public Health–Seattle and King County. 2012a. Snow and Ice Storm: After Action Report, January 17-21. Available at http://www.seattle.gov/emergency/publications/documents/AAR2012January16-19SnowFinal.pdf.

Public Health–Seattle and King County. 2012b. Vulnerable Populations Action Team (VPAT). Available at http://www.kingcounty.gov/healthservices/health/preparedness/VPAT.aspx.

Quinn, S. C., S. Kumar, V. S. Freimuth, D. Musa, N. Casteneda-Angarita, and K. Kidwell. 2011. Racial disparities in exposure, susceptibility, and access to health care in the US H1N1 influenza pandemic. *American Journal of Public Health* 101(2):285-293.

Roussos, S. T., and S. B. Fawcett. 2000. A review of collaborative partnerships as a strategy for improving community health. *Annual Review of Public Health* 21:369-402.

Ryan, J., and J. Hawdon. 2008. From individual to community: The "framing" of 4-16 and the display of social solidarity. *Traumatology* 14(1):43-51.

Satcher, D. 2011. The impact of disparities in health on pandemic preparedness. *Journal of Health Care for the Poor and Underinsured* 22:36-37.

Schoch-Spana, M., B. Courtney, C. Franco, A. Norwood, and J. B. Nuzzo. 2008. Community resilience roundtable on the implementation of Homeland Security Presidential Directive 21 (HSPD-21). *Biosecurity and Bioterrorism* 6(3):269-278.

Schoch-Spana, M., C. Franco, J. B. Nuzzo, and C. Usenza. 2007. Community engagement: Leadership tool for catastrophic health events. *Biosecurity and Bioterrorism: Biodefense Strategy, Practice, and Science* 5(1):8-25.

Sonn, C. C., and A. T. Fisher. 1998. Sense of community: Community resilient responses to oppression and change. *Journal of Community Psychology* 26(5):457-472.

Spence, P. R., K. A. Lachlan, and D. R. Griffin. 2007. Crisis communication, race, and natural disasters. *Journal of Black Studies* 37(4):539-554.

Tierney, K. J. 2000. Executive Summary: Disaster Resistant Communities Initiative: Evaluation of the Pilot Phase, Year 2. Submitted to the Federal Emergency Management Agency by the University of Delaware Disaster Research Center. Available at http://www.udel.edu/DRC/archives/Documents/Project%20Impact/projectreport42.pdf.

Tierney, K., M. K. Lindell, and R. W. Perry. 2001. *Facing the Unexpected: Disaster Preparedness and Response in the United States.* Washington, DC: The National Academies Press.

Timm, K. L. 2004. Code enforcement consolidation: Is it right for the City of Two Rivers? Submitted to the National Fire Academy as part of the Executive Fire Officer Program. Available at http://www.usfa.fema.gov/pdf/efop/efo37649.pdf.

Toner, E., R. Waldhorn, and B. Courtney. 2009. Hospitals Rising to the Challenge: The First Five years of the U.S. Hospital Preparedness Program and Priorities Going Forward. Prepared by the Center for Biosecurity of UPMC for the U.S. Department of Health and Human Services.

Torrey, C., M. Burke, M. Lee, A. Dey, S. Fussell, and S. Kiesler. 2007. Connected giving: Ordinary people coordinating disaster relief on the Internet. Human-Computer Interaction Institute, Carnegie Mellon University, paper 51. Available at http://www.computer.org/portal/web/csdl/doi?doc=doi/10.1109/HICSS.2007.144

Trust for America's Health. 2011. *Ready or Not? Protecting the Public's Health from Diseases, Disasters and Bioterrorism*. Washington, DC: TFAH. Available at http://healthyamericans.org/assets/files/TFAH2011ReadyorNot_09.pdf. Accessed June 17, 2012.

U.S. Bureau of Labor Statistics. 2009. Occupational Outlook Handbook, 2010-11 Ed., Construction and Building Inspectors. Available at http://www.bls.gov/oco/ocos004.htmhttp://www.bls.gov/oco/ocos004.htm.

U.S. Census Bureau. 2010. 2010 Census. Available at http://www.census.gov/popest/cities/.

USGS (U.S. Geological Survey). 2008. The ShakeOut Scenario. USGS Open File Report 2008-1150. Available at http://pubs.usgs.gov/of/2008/1150/of2008-1150small.pdf.

Vaughn, E. 1995. The significance of socioeconomics and ethnic diversity for the risk communication process. *Risk Analysis* 15(2):169-180.

Wachtendorf, T., and J. M. Kendra. 2004. Considering Convergence, Coordination, and Social Capital in Disasters. University of Delaware Disaster Research Center Preliminary Paper No. 342a. Available at http://dspace.udel.edu:8080/dspace/bitstream/handle/19716/737/PP342A.pdf?sequence=1 .

Waugh, W. L., and G. Streib. 2006. Collaboration and leadership for effective emergency management. *Public Administration Review* 66:131-140.

Weinstein, N. D., and M. Nicolich. 1993. Correct and incorrect interpretations of correlations between risk perceptions and risk behaviors. *Health Psychology* 12(3):235-245.

Wood, M. M., D. S. Mileti, M. Kano, M. M. Kelley, R. Regan, and L. B. Bourque. 2011. Communicating actionable risk for terrorism and other hazards. *Risk Analysis* 32(4):601-615.

Zoraster, R. M. 2010. Vulnerable populations: Hurricane Katrina as a case study. *Prehospital and Disaster Medicine* 25(1):74-78.

"Resilience begins with leadership, appropriate planning both in terms of action-plans but also in terms of proper community planning and development visions."
Dr. Larry Weber, University of Iowa

6

The Landscape of Resilience Policy—Resilience from the Top Down

INTRODUCTION

The key elements of resilience include strong governance at all levels, including the making of consistent and complementary local, state, and federal policies. As previously discussed, communities are not under a single authority, but must function under a mix of policies and practices implemented and enforced by different levels of government. Furthermore, policies that make the nation more resilient are important in every aspect of American life and economy, and not just during times of stress or trauma. A key role of policies designed to improve national resilience is to take the long-term view of community resilience and to help avoid short-term expediencies that can diminish resilience. Policies to improve community and national resilience may be designed and promulgated specifically to address issues of resilience, or they may be policies designed for another reason that acknowledge the importance and process of building resilience. In some cases, policies designed to accomplish one positive goal may unintentionally cause deterioration of community resilience. Therefore, policies and programs at all levels of government require examination to assess their impact on the long-term resilience of communities and the nation.

Increasing national resilience through specific policy measures involves addressing the multiple aspects of resilience that have been discussed in this report. For example, as Chapter 2 emphasizes, policy mechanisms play a role in risk management through provision of data and information to evaluate potential hazards, although, as Chapter 2 outlined, information alone does not ensure resilience. Likewise, progress toward improved resilience is driven by the need and value propositions outlined in Chapter 3, and likely monitored using the indicators and tools described in Chapter 4 of this report. At the national level, policies that enhance national resilience are not simply disaster reduction policies. Because the scope of resilience is sometimes not fully appreciated, some who contemplate national resilience policy think first of the Stafford Act and its role in disaster response and recovery. Although the Stafford Act (discussed further below) does provide for certain responsibilities and actions in the face of a disaster, national resilience, as has been demonstrated throughout this report, transcends the immediate impact and disaster response and, therefore, grows from a broader set of policies. Many of the policies that affect national resilience are not related to specific hazards or disaster events at all, including some policies that may apply only to specific subsystems of a community (Longstaff et al., 2010), and others that may have effects on essential community services such as education and health care (see Chapter 5).

With this background, this chapter is developed from the idea that improvement of national resilience relies on collections of coordinated and integrated policies at multiple levels rather than a single comprehensive government policy. The subsequent sections provide context for considering policy options across the full range of stakeholders and authorities that constitute the landscape of resilience, and describes several current practices at federal, state, and local levels that support resilience, as well as policies that unintentionally undermine resilience. Identification of specific roles and responsibilities of government in building resilience flows naturally from discussion in Chapter 5 of the complementary roles and actions that communities can embrace as part of a systemic national effort to increase resilience. The interdependency and interaction of community initiatives and government policy are critical for increasing resilience (see Chapter 7 for the way in which bottom-up and top-down approaches may be linked).

EXISTING FEDERAL POLICIES THAT STRENGTHEN RESILIENCE

Federal policies are intended to provide a set of nationally uniform laws or practices to address national needs that transcend the conditions or needs of individual states or cities. Federal policies address issues that have national scope and importance, even if the issues and consequences are local. These policies exist at the level of the Executive Branch—in both the Office of the President and in the Cabinet Departments as well as in independent federal agencies—and in laws enacted by the Legislative Branch. An outline of the most critical of the policies that the committee determined would provide support to strengthen resilience is briefly reviewed below.

Federal Executive Branch Policies Supporting Resilience

U.S. national leaders continue to seek broad policies for strengthening the nation against both terrorist acts and natural disasters. Certain Executive Branch policies, for example, are promulgated by the President through Executive Orders or Directives that guide the actions of federal agencies. These Presidential Directives and Executive Orders have the force of law. Directives may take different forms, but most recent Presidential Directives affecting national resilience have been either Homeland Security Presidential Directives (HSPD) or Presidential Policy Directives (PPD). A Presidential Policy Directive (PPD-8) from 2011 entitled "National Preparedness" begins by saying:

> *This directive is aimed at strengthening the security and resilience of the United States through systematic preparation for the threats that pose the greatest risk to the security of the Nation, including acts of terrorism, cyberattacks, pandemics, and catastrophic natural disasters.* (White House and DHS, 2011)

The Directive calls for the development of a National Preparedness System to guide activities that will enable the nation to achieve the goal of strengthening its security and resilience; for a comprehensive campaign to build and sustain national preparedness; and for an annual National Preparedness Report to measure progress in meeting the goal. Importantly, the President calls on DHS to embrace systematic preparation against all types of threats, including catastrophic natural disasters.

Preparedness is not synonymous with resilience, but they are related. According to PPD-8, "The term 'resilience' refers to the ability to adapt to changing conditions and withstand and rapidly recover from disruption due to emergencies" (White House and DHS, 2011). This definition is in keeping with the definition of resilience established by the committee during the course of this study (see Chapter 1). The Directive also recognizes resilience as a characteristic of an individual, community, or nation and that resilience is enhanced through improved preparedness as noted below:

> *The Secretary of Homeland Security shall coordinate a comprehensive campaign to build and sustain national preparedness, including public outreach and community-based and private-sector programs to enhance national resilience, the provision of Federal financial assistance, preparedness efforts by the Federal Government, and national research and development efforts.* (White House and DHS, 2011)

As Box 6.1 shows, an entire series of HSPDs has been issued since September 11, 2001. Although many of these directives are heavily focused on terrorist threats, the preparation and response of communities to terrorist threats contain many of the same elements as preparation for natural hazards. Thus, significant and deliberate overlap exists in the application of HSPDs to both human-made and natural threats. PPD-8 is one that can be broadly applied in this way.

Importantly, PPD-8 recognizes that our national response to a wide range of events, from the 2009 H1N1 pandemic to the BP Deepwater Horizon oil spill, has been strengthened by leveraging the expertise and resources that exist in our communities. The Department of Homeland Security (DHS) is directed to coordinate a "comprehensive campaign," informed by the long-term requirements for national resilience, to reach the goals of the Directive. Although the President assigns the Secretary of DHS to coordinate this comprehensive campaign under PPD-8, the directive indicates that DHS is not expected to conduct all of the work itself, but to coordinate the work of others. The Committee supports the role of DHS in serving as coordinator of these broad efforts to enhance national resilience under PPD-8 (see additional discussion in Chapter 7).

BOX 6.1
Homeland Security Presidential Directives Relevant to National Resilience

- HSPD-1: Organization and Operation of the Homeland Security Council. Ensures coordination of all homeland security-related activities among executive departments and agencies and promotes the effective development and implementation of all homeland security policies (October 2001).
- HSPD-3: Homeland Security Advisory System. Establishes a comprehensive and effective means to disseminate information regarding the risk of terrorist acts to federal, state, and local authorities and to the American people (March 2002). This system was replaced by the Terrorism Advisory System in 2011.
- HSPD-5: Management of Domestic Incidents. Enhances the ability of the United States to manage domestic incidents by establishing a single, comprehensive national incident management system (February 2003).
- HSPD-7: Critical Infrastructure Identification, Prioritization, and Protection. Establishes a national policy for federal departments and agencies to identify and prioritize U.S. critical infrastructure and key resources and to protect them from terrorist attacks (December 2003).
- HSPD-8 Annex 1: National Planning. Rescinded by PPD-8 (below): National Preparedness, except for paragraph 44. Individual plans developed under HSPD-8 and Annex I remain in effect until rescinded or otherwise replaced (December 2003).
- Presidential Policy Directive/PPD-8: National Preparedness. Aimed at strengthening the security and resilience of the United States through systematic preparation for the threats that pose the greatest risk to the security of the nation, including acts of terrorism, cyber attacks, pandemics, and catastrophic natural disasters (March 2011).
- HSPD-20: National Continuity Policy. Establishes a comprehensive national policy on the continuity of federal government structures and operations and a single national continuity coordinator responsible for coordinating the development and implementation of federal continuity policies (May 2007).

- HSPD-20 Annex A: Continuity Planning. Assigns executive departments and agencies to a category commensurate with their COOP/COG/ECG responsibilities during an emergency (September 2008).
- HSPD-21: Public Health and Medical Preparedness. Establishes a national strategy that will enable a level of public health and medical preparedness sufficient to address a range of possible disasters (October 2007).
- HSPD-23: National Cyber Security Initiative (January 2008).

Source: DHS, http://www.dhs.gov/xabout/laws/editorial_0607.shtm.
Notes: PPD-8 (http://www.dhs.gov/xabout/laws/gc_1215444247124.shtm) replaces HSPD-8 (2003) and HSPD-8 Annex I (2007). Relevance of all HSPDs in this list to national resilience has been evaluated by the Committee for this study.

The language of PPD-8 makes clear that American communities and the private sector play central roles in enhancing national resilience and, therefore, that DHS's coordination of federal efforts also involves effective engagement of those critical stakeholders. Significantly, DHS is also called upon to coordinate federal financial assistance, the preparedness efforts by other federal agencies, and national research and development efforts.

The issuance of PPD-8 was a significant advance in increasing and improving the federal role in national resilience, and its goals were amplified by the report of the Homeland Security Advisory Council's Community Resilience Task Force (CRTF, 2011). That report, released in June 2011, builds on the Quadrennial Homeland Security Review Report[1] and contains a set of recommendations intended to define the role of DHS in advancing national resilience through the mechanism of PPD-8:

> *The Department of Homeland Security (DHS) clearly has an important role to play in building national resilience, but at its core, the resilience charge is about enabling and mobilizing American communities. The CRTF acknowledges that many relevant activities are already underway, particularly in fostering development of preparedness capabilities, but observes that those activities are rarely linked explicitly to resilience.* (CRTF, 2011)

---

[1]The Quadrennial Homeland Security Review Report (http://www.dhs.gov/xlibrary/assets/qhsr_report.pdf) contains five Homeland Security missions. Mission 5 is Resilience to Natural Disasters, which outlines the traditional elements of hazard mitigation, enhanced preparedness, effective emergency response, and rapid recovery. These issues are also discussed in the DHS Bottom-Up Review Report (http://www.dhs.gov/xlibrary/assets/bur_bottom_up_review.pdf) released in July, 2010.

The recommendations contained in the CRTF report (Box 6.2) represent a strong and clear starting point for federal involvement in building national resilience. The recommendations are directed specifically to DHS and call for clarification of responsibilities, building knowledge and public awareness to enhance individual and societal resilience, and providing long-term targets to support urban planning and the built environment.

BOX 6.2
Recommendations of the Homeland Security Advisory Council, Community Resilience Task Force (CRTF)
2011

CRTF Recommendations that apply across the full range of Community Resilience activities include:

CRTF Recommendation 1.1: *Build a Shared Understanding of the Shared Responsibility.* DHS should take the lead in working with key stakeholder groups to develop and share models for resilience—illustrations of resilience in operational settings—within the context of each group. The purpose is to motivate stakeholders to learn from each other and to do what they can to enhance resilience without waiting for external intervention.
CRTF Recommendation 1.2: *Build a Coherent and Synergistic Campaign to Strengthen and Sustain National Resilience.* DHS should align policies, programs, and investments to motivate and operationalize resilience, and should use its leadership charge from PPD-8 to motivate similar actions across the federal government and throughout the Nation.
CRTF Recommendations 1.3: *Organize for Effective Execution.* DHS should establish a National Resilience Office and charge it with building the resilience foundation envisioned by the QHSR.
CRTF Recommendation 1.4: *Build the Knowledge and Talent Base for Resilience.* DHS should implement a research program to build the intellectual underpinnings for resilience training and education programs to be delivered throughout the Nation.

CRTF Recommendations related to enhancing individual and societal resilience include:

CRTF Recommendation 2.1: *Update ready.gov.* DHS should establish and execute a plan for periodic review and update of the content and presentation of information on ready.gov; messages should be linked explicitly to resilience outcomes.
CRTF Recommendation 2.2: *Build Public Awareness.* DHS should develop and implement a comprehensive and coherent suite of communications strategies in support of a national campaign to increase public awareness and motivate individual citizens to build societal resilience.

CRTF Recommendation 2.3: *Motivate and Enable Action.* DHS should adapt and implement proven incentive and award programs to motivate individual and community engagement and action, and further develop mechanisms to facilitate and enable engagement.

CRTF Recommendations targeting urban planning for the built environment include:

CRTF Recommendation 3.1: *Leverage Existing Federal Assets.* DHS, in conjunction with the General Services Administration and local officials, should develop a Resilient Community Initiative (RCI) that leverages federal assets and programs to enable community resilience.
CRTF Recommendation 3.2: *Align Federal Grant Programs to Promote and Enable Resilience Initiatives.* DHS should review and align all grant programs related to infrastructure or capacity building, and should support development of synchronized strategic master plans for improvement of operational resilience throughout the Nation.
CRTF Recommendation 3.3: *Enable Community-Based Resilient Infrastructure Initiatives.* DHS should transform its critical infrastructure planning approach to more effectively enable and facilitate communities in their efforts to build and sustain resilient critical infrastructures.
CRTF Recommendation 3.4: *Enable Community-Based Resilience Assessment.* DHS should coordinate development of a community-based, all-hazards American Resilience Assessment (ARA) methodology and toolkit.

Source: Homeland Security Advisory Council, Community Resilience Task Force (http://www.dhs.gov/xlibrary/assets/hsac-community-resilience-task-force-recommendations-072011.pdf), June 2011.

In addition to the CRTF recommendations, the National Preparedness Goal developed by DHS in response to PPD-8 provides a statement of national preparedness that includes preemptive actions designed to mitigate or reduce the impact of both terrorism and natural hazards in order to develop a more resilient nation (Box 6.3). The National Preparedness Goal deals with preparedness across jurisdictions and at a national scale.

The formulation of the National Preparedness Goal, the operational implementation of its many aspects, and the administration of several community funding programs, primarily through the Federal Emergency Management Agency (FEMA),[2] place DHS in a strong position to provide leadership in the interagency efforts required to build national resilience.

[2] http://www.dhs.gov/ynews/releases/20110217-dhs-fy12-grant-guidance.shtm.

BOX 6.3
DHS National Preparedness Goal (excerpt)

"We describe our security and resilience posture through the core capabilities . . . that are necessary to deal with great risks, and we will use an integrated, layered, and all-of-Nation approach as our foundation. We define success as:

A secure and resilient Nation with the capabilities required across the whole community to prevent, protect against, mitigate, respond to, and recover from the threats and hazards that pose the greatest risk.

Using the core capabilities, we achieve the National Preparedness Goal by:

– Preventing, avoiding, or stopping a threatened or an actual act of terrorism.
– Protecting our citizens, residents, visitors, and assets against the greatest threats and hazards in a manner that allows our interests, aspirations, and way of life to thrive.
– Mitigating the loss of life and property by lessening the impact of future disasters.
– Responding quickly to save lives, protect property and the environment, and meet basic human needs in the aftermath of a catastrophic incident.
– Recovering through a focus on the timely restoration, strengthening, and revitalization of infrastructure, housing, and a sustainable economy, as well as the health, social, cultural, historic, and environmental fabric of communities affected by a catastrophic incident.

...These are not targets for any single jurisdiction or agency; achieving these targets will require a national effort involving the whole community."

Source: Department of Homeland Security, National Preparedness Goal, 1st Edition, September, 2011, http://www.fema.gov/pdf/prepared/npg.pdf.

The conduct of federal activities in partnership with state, local, and private partners may also be the goal of other Presidential directives. For example, the interaction of federal agencies with the private sector to advance the goal of improving resilience has been demonstrated in the area of critical infrastructure. Homeland Security Presidential Directive 7 (HSPD-7) gives the Secretary of Homeland Security oversight responsibility for protecting 18 critical infrastructure sectors, and gives selected agencies and the Environmental Protection Agency the ability to direct national infrastructure protection for some sectors (Box 6.4). These responsibilities require close coordination with state and local government, as well as the private sector, and may provide a model for the federal–state–local–private partnerships required to develop broader strategies for building resilience in U.S. communities.

BOX 6. 4
Roles and Responsibilities of Sector-Specific Federal Agencies in Critical Infrastructure Protection

"18. Recognizing that each infrastructure sector possesses its own unique characteristics and operating models, there are designated Sector-Specific Agencies, including

a. Department of Agriculture—agriculture, food (meat, poultry, egg products);
b. Health and Human Services—public health, health care, and food (other than meat, poultry, egg products);
c. Environmental Protection Agency—drinking water and water treatment systems;
d. Department of Energy—energy, including the production refining, storage, and distribution of oil and gas, and electric power except for commercial nuclear power facilities;
e. Department of the Treasury—banking and finance;
f. Department of the Interior—national monuments and icons; and
g. Department of Defense—defense industrial base.

19. In accordance with guidance provided by the Secretary, Sector-Specific Agencies shall:

a. collaborate with all relevant Federal departments and agencies, State and local governments, and the private sector, including with key persons and entities in their infrastructure sector;
b. conduct or facilitate vulnerability assessments of the sector; and
c. encourage risk management strategies to protect against and mitigate the effects of attacks against critical infrastructure and key resources."

Source: Homeland Security Presidential Directive 7: Critical Infrastructure Identification, Prioritization, and Protection, December 17, 2003.

Other types of federal policies may also strongly affect resilience in very broad ways. For example, evidence is growing that changing global climate is increasing the nation's exposure to natural hazards through more frequent and severe storms, as well as more extensive droughts and increased vulnerability of our coastal regions through sea-level rise (NRC, 2012). Thus, one type of long-term federal policy goal to improve U.S. national resilience might include an energy policy that addresses carbon emissions and dependence on imported energy resources. Addressing carbon emissions could help mitigate climate change which otherwise may result in an increase in frequency and intensity of weather-related hazards and could help support a national effort to become less import-dependent for our energy needs (NRC, 2010). Although such policies may not be recognized immediately as affecting resilience to natural disasters, they are examples of the far-reaching implications of policy decisions that may have impact on national resilience.

Finally, strategic investment of federal funds in local communities—even within the structure of existing statutes and programs—may provide a strong impetus to develop more resilient communities. Communities realize that stronger infrastructure and institutions would make their population less vulnerable to disasters, but they generally lack the resources or political will to make capital-intense short-term investments even if they believe that those investments will reap long-term benefits. In the future, predisaster funding may serve as a critical tool in building national resilience. The practice of federal funding of post-disaster recovery within local communities should be strategically complemented with predisaster funding of the highest-priority resilience elements within a community, such as enforcement of building codes, land-use and development planning, and disaster-resistant health care services. Existing programs such as those within FEMA[3] could be strengthened to place a greater emphasis on resilience. Careful analysis and consideration of a strategic approach to federal funding of resilience are important in efforts to reduce the impact (and cost) of disasters.

Coordination of Executive Branch Federal Agencies

In addition to the Executive Branch policies issued through Presidential Directives and Executive Orders, agency policies may be initiated by individual federal agencies through the rulemaking process, and may include such things as management practices for federal lands or other resources, or rules and policies that outline roles and responsibilities of various federal agencies in managing federal assets, including those directing or supporting the activities that foster community resilience. A key challenge for the federal government is how to maintain motivation and accountability among all of the federal agencies in the pursuit of defined, common goals toward increasing resilience. Each federal agency has a specific mission, has a budget that is largely separate from the budgets of other agencies, and is accountable to the President and to Congress, rather than to other agencies.

A large number of federal agencies play key roles in mitigation, preparedness and response aspects of building resilience. The ways in which federal agencies are coordinated to address resilience issues on individual, community, state, and national levels are currently not always clear, and the process of coordination should be defined around a common vision of resilience in order to leverage the effectiveness of each agency's efforts and investments. DHS, by virtue of its mission and because it contains the major response

[3] www.fema.gov/government/grant/hma/index.shtm.

agencies, FEMA and the Coast Guard, houses much of the federal responsibility and accountability for fostering national resilience and has a leading role during response to incidents. However, DHS partners with other agencies that provide research, information, and response capabilities essential to national resilience. The National Oceanic and Atmospheric Administration (NOAA), the U.S. Geological Survey (USGS), the National Aeronautics and Space Administration, the U.S. Forest Service, and the the U.S. Army Corps of Engineers play crucial roles in providing scientific understanding and real-time assessments of weather-related issues, fires, earthquakes, floods, tornadoes, and other natural hazards, relevant both for short- and long-term monitoring and planning before disasters occur and during actual events. The Corps of Engineers, the Bureau of Reclamation, the National Resources Conservation Service, and the Federal Energy Regulatory Commission manage or provide oversight for levees and other structures and therefore play a critical role in flood reduction and management, water supply, and energy generation. The Department of Energy has key responsibilities for the energy infrastructure—coordinating such aspects as energy infrastructure security and energy restoration, and emergency preparedness and response for critical energy infrastructure.

In addition to attention to natural science and infrastructure components, resilience relies on the health and welfare of the citizenry, and so federal agencies such as the Department of Health and Human Services, the Department of Education, and the Department of Housing and Urban Development, and other federal agencies play key roles in helping to build the total resilience of U.S. communities. A partial list of the numerous federal departments and agencies engaged in some aspect of building community and national resilience is shown in Table 6.1 along with some of their ongoing resilience-related activities and initiatives. Of course it is difficult to coordinate these numerous and diverse federal efforts, but failure to adequately harmonize the work of these agencies reduces the effectiveness of the overall federal effort to increase national resilience. On the other hand, improved coordination of federal resilience programs in communities provides significant opportunities for leveraging federal funding and ensuring that agencies are not working at cross purposes.

Many agencies have demonstrated successful federal–state–local–private cooperation arising from internal agency vision or goals, For example, USGS and NOAA have worked with nonfederal partners to transfer research results to their stakeholders, and have worked successfully to help communities to assess and mitigate their earthquake and coastal hazards. These successful examples have not happened by accident, but result from explicit policies within each agency. The vision statement from the NOAA Administrator in the agency's 5-year plan says:

> NOAA's mission is central to many of today's greatest challenges. The state of the economy. Jobs. Climate Change. Severe weather. Ocean acidification. Natural and human-

> induced disasters. Declining biodiversity. Threatened or degraded oceans and coasts. These challenges convey a common message: Human health, prosperity, and well-being depend upon the health and resilience of both managed and unmanaged ecosystems. Combined with the capabilities of our many partners in government, universities, and private and nonprofit sectors, NOAA's science, service, and stewardship capabilities can help transition to a future where societies and world's ecosystems reinforce each other and are mutually resilient in the face of sudden and prolonged change. (NOAA, 2012)

And the USGS states:

> The USGS brings the results of its many research programs together to create knowledge that is understandable, useable, and accessible in many forms—including statistics, reports, analyses, maps, models, and tools that forecast the consequences of various choices. These products, often created in partnership with other governmental, academic, and private organizations, provide the basis for evaluating the effectiveness of specific policies and management actions, and they are essential to the success of policymakers and decisionmakers at local, State, Federal, tribal, and international levels. (USGS, 2009)

Despite the intent behind written statements such as the examples above, coordination of federal agencies' efforts to promote and build national resilience will be difficult owing to the independence of federal agencies, each with its own mission and budget and each emphasizing disaster planning, homeland security, or resilience to different degrees. However, no consistently owned and applied vision for national resilience can exist without coordination of federal agencies. Interagency coordination is essential to a number of other federal efforts, and many interagency coordination groups already exist with varying degrees of effectiveness. To work effectively and to ensure participation by all key agencies, such an interagency working group would necessarily be convened or created and charged by the Executive Office or Congress. Coordinating investments among federal agencies is exceedingly difficult, but a common vision of national resilience developed with the participation of all key federal agencies, and with input from state, local, and private-sector stakeholders would improve the consistency with which those funds are applied.

As discussed above, PPD-8 provides clear presidential direction for coordination of federal efforts to enhance national resilience, and coordination of policies and procedures among federal agencies are further discussed in Chapter 7.

TABLE 6.1 Examples of Federal Efforts Among the Study's Sponsors and Other Federal Agencies That Contribute to Enhanced Disaster Resilience

Federal Departments and Agencies	Ongoing or Planned Goal, Program, Project, or Initiative	Web Link
U.S. Department of Agriculture (USDA)	*Goals 1 and 2 of USDA Strategic Plan 2010-2015*: 1) Assist rural communities, including expansion of USDA "work with landowners to increase adoption of practices that will make farms, ranches, and forestlands more *resilient* to the effects of climate change." (p. 3) 2) "Ensure that our national forests and private working lands are conserved, restored, and made more *resilient* to climate change, while enhancing our water resources." (p. 14)	http://www.ocfo.usda.gov/usdasp/sp2010/sp2010.pdf (in addition to the U.S. Forest Service, other offices in the USDA with resilience-oriented initiatives include the National Institute of Food and Agriculture and the Agricultural Research Service
U.S. Forest Service	*National Roadmap for Responding to Climate Change*, including a *Performance Scorecard*: The Roadmap describes agency response to climate change through "adaptive restoration—by restoring the functions and processes characteristic of healthy, *resilient* ecosystems" (p. 18). The Scorecard includes elements about organization, leadership, partnerships, adaptation, mitigation, and sustainability.	(1) http://www.fs.fed.us/climatechange/pdf/Roadmapfinal.pdf (2) http://www.fs.fed.us/climatechange/advisor/scorecard.html

U.S. Department of Commerce		
National Oceanic and Atmospheric Administration	*National Ocean Service/Coastal Services Center and Office of Ocean and Coastal Resource Management*: Coastal resilience initiatives to allow communities to "bounce back" after a disaster; programs, data, tools, analysis, projects, and training allow users (coastal management community) access to information important for coastal resilience. A guide to coastal community resilience outlines coastal hazards, the importance of coastal resilience, and steps for coastal communities to take to become more resilient and to assess progress. *NOAA and Mississippi-Alabama Sea Grant Consortium*: Coastal Resiliency Index: A community self-assessment aims to provide community leaders with straightforward, inexpensive ways to gauge whether their community will return to a satisfactory level of functioning after a disaster; in other words, to allow communities to measure their progress toward becoming disaster resilient. (Link 4) *Disaster Resilient Communities: A NIST/NOAA Partnership*: See description under National Institute of Standards and Technology, below. (Link 5) *National Weather Service*: Provides local and regional data and forecasts regarding weather situations (e.g., storms, hurricanes, floods, tornadoes, tsunamis) (Links 6, 7)	(1) http://oceanservice.noaa.gov/facts/resilience.html (2) http://www.csc.noaa.gov/psc/riskmgmt/resilience.html (3) http://www.crc.uri.edu/download/CCRGuide_lowres.pdf (4) http://masgc.org/page.asp?id=591 (5) http://fire.nist.gov/bfrlpubs/build07/PDF/b07037.pdf (6) http://www.nws.noaa.gov/ (7) http://www.nws.noaa.gov/com/weatherreadynation/

National Institute of Standards and Technology	*Disaster* ***Resilient*** *Communities: A NIST/NOAA Partnership*: A plan that addresses wildland fires, the effects of winds (hurricanes, tornadoes, and other winds), storm surge, tsunamis, and earthquakes, particularly on coastal communities.	http://fire.nist.gov/bfrlpubs/build07/PDF/b07037.pdf
Engineering Laboratory	*Disaster-resilient buildings, infrastructure, and communities*: Developing and applying measurement methods, models, and tools to reduce risk and increase *resilience* of buildings, infrastructure, and communities. Related areas include earthquake and fire risk reduction for buildings and communities, windstorm impact reduction, and behavior of structures under multihazard situations.	http://www.nist.gov/el/disresgoal.cfm
U.S. Department of Defense (DOD)	*Defense Critical Infrastructure Program (DCIP)*: DOD role described in the overarching document Homeland Security Presidential Directive 7	HSPD-7: http://www.dhs.gov/xabout/laws/editorial_0607.shtm (in addition to the U.S. Army Corps of Engineers, other elements of the DOD involved directly in resilience-related activities include the National Guard Bureau and the U.S. Northern Command)
U.S. Army Corps of Engineers (USACE)	*Public works defense infrastructure (related to the DCIP):* Critical infrastructure protection and resilience. All-hazards approach. Has involved several regional resilience studies of dams and watersheds. *Civil works:* flood damage reduction, water and power	(1) http://www.dtic.mil/ndia/2010homeland/Seda_Sanabria.pdf (2) http://www.usace.army.mil/Locations.aspx (3)

	supply, regulatory program (for U.S. waters), and protection of resources. District offices with direct responsibility and oversight. (Link 2) *Emergency response (includes disaster response, flood control and coastal emergencies, emergency support*: Responds in several ways as part of federal government's unified national response to disasters; activities include providing engineering expertise to local and state governments, providing essential resources such as drinking water, auxiliary power, temporary housing and roofing, making repairs to critical infrastructure. Emergency support function assists other federal agencies, particularly DHS and FEMA, and is performed in concert with federal, state, and local governments, contractors, and industries. Supports DHS disaster response framework. (Link 3)	http://www.usace.army.mil/Media/FactSheets/FactSheetArticleView/tabid/219/Article/156/emergency-response.aspx
U.S. Navy	The U.S. Navy Climate Change Roadmap addresses the national security issues associated with climate change. The roadmap presents the ways in which the U.S. Navy will observe, predict, and adapt to climate change.	http://www.navy.mil/navydata/documents/CCR.pdf
U.S. Department of Energy		
Office of Infrastructure Security and Energy Restoration	Coordinates DOE's response to energy emergencies, contributes to the security of the national energy infrastructure, aids local and state governments with planning, preparation, and response to energy emergencies	http://energy.gov/oe/mission/infrastructure-security-and-energy-restoration-iser

U.S. Department of Health and Human Services		
Centers for Disease Control and Prevention	Health protection agency for the nation; works to protect people from public health threats, including bioterrorism, chemical and radiation emergencies, disease outbreaks, and medical emergencies arising from natural disasters. CDC's Office of Public Health Preparedness and Response leads the agency's preparedness and response activities by providing strategic direction, support, and coordination for activities across CDC as well as with local, state, tribal, national, territorial, and international public health partners. CDC provides funding and technical assistance to states to build and strengthen public health capabilities. Ensuring that states can adequately respond to threats will result in greater health security.	http://www.cdc.gov/ http://www.cdc.gov/phpr/http://emergency.cdc.gov/
Office of the Assistant Secretary for Preparedness and Response (ASPR)	National and community preparation to respond to and recover from public health and medical disasters and emergencies. Key goals include promoting resilient communities, strengthening federal public health and medical leadership, promoting effective countermeasures, improving health care delivery systems, strengthening ASPR leadership and management. Office of Preparedness and Emergency Operations under ASPR has responsibility for operational plans, tools, and training to ensure response and recovery from health and medical emergencies. Coordinates with other federal agencies during	http://www.phe.gov/about/aspr/strategic-plan/Pages/default.aspx

	emergencies.	
Office of the National Coordinator for Health Information Technology	Tasked with creating a national database and a plan for "the utilization of an electronic health record (EHR) for each person in the United States by 2014." These records can allow access to key medical data for those affected by disasters who are in need of treatment and medications.	http://healthit.hhs.gov/portal/server.pt/community/healthit_hhs_gov__home/1204
U.S. Department of Homeland Security (DHS)	DHS works to build a resilient nation. The agency provides the coordinated federal response to events such as terrorist attacks, natural disasters or other large-scale emergencies. DHS also works with federal, state, local and private sector partners in recovery efforts.	http://www.dhs.gov/building-resilient-nation
Federal Emergency Management Agency (FEMA)	FEMA supports the public and first responders to prepare for, protect against, respond to, recover from, and mitigate all hazards, FEMA's statutory authority derives from the Stafford Act (P.L. 100-707) (Link 1) Guiding documents and plans include the National Response Framework (Link 2), a "Whole Community" operational approach (Link 3), and a National Disaster Recovery Framework (Link 4).	(1) http://www.fema.gov/about/index.shtm (2) http://www.fema.gov/emergency/nrf/ (3) http://www.fema.gov/about/wholecommunity.shtm (4) http://www.fema.gov/national-disaster-recovery-framework
Science and Technology Directorate	Manages science and technology research for homeland protection, including development and transition for use by first responders. Research includes physical and engineering science as well as human factors and behavioral science.	http://www.dhs.gov/xabout/structure/editorial_0530.shtm
U.S. Department of Housing and Urban Development		

Community Planning and Development Program	The office works to develop viable communities that provide opportunities for low- and moderate-income people through a variety of public–private partnerships. Aspects of these efforts include affordable housing as well as community and economic development.	http://portal.hud.gov/hudportal/HUD?src=/program_offices/comm_planning
Office of Public and Indian Housing	The office focuses on safe, affordable housing including creation of possibilities for residents to become self-sufficient and economically stable (Link 1). The Capital Fund Emergency/Natural Disaster Funding is a financial reserve for public housing agencies (PHAs) that experience emergency situations or a natural disaster subject to compliance with certain requirements (Link 2).	(1) http://portal.hud.gov/hudportal/HUD?src=/program_offices/public_indian_housing (2) http://portal.hud.gov/hudportal/HUD?src=/program_offices/public_indian_housing/programs/ph/capfund/emfunding
U.S. Department of the Interior		
U.S. Geological Survey	Primary earth science organization in the Department of the Interior. As part of the agency's purview over processes operating in the Earth science system, it includes a specific focus on monitoring and assessing natural hazards and helping to develop strategies for resilience.	http://www.usgs.gov/natural_hazards/; http://pubs.usgs.gov/fs/2011/3008/
Independent Agencies and Corporations:		
National Aeronautics and Space Administration		
Applied Sciences	The program uses Earth science data derived from NASA	http://www.coastal.ssc.nasa.gov/

Program	research to address a variety of topics including coastal community *resilience*.	
Oak Ridge National Laboratory (ORNL)/Community and Regional Resilience Institute (CARRI)	Originally housed at the ORNL, CARRI is now housed at the Meridian Institute. The CARRI mission aims to help develop and share information and guidance that communities may use to prepare for, respond to, and rapidly recover from human-made or natural disasters with minimal downtime of basic services.	http://www.resilientus.org/

Note: This table is not an exhaustive list of all federal programs or activities that may or could contribute to increasing national disaster resilience.

Federal Legislation

Communities across the nation rely on federal policies that help advance resilience. Congress and other policymakers can improve the resilience of communities and the nation by taking a holistic view of the diverse aspects of community resilience when developing policies of all kinds as well as recognizing the complex interactions of specific federal policies with each other and their likely effect on the communities themselves.

Legislative Branch policies may be established and implemented explicitly through legislation, or implicitly through the oversight process that holds federal agencies accountable through the hearings or appropriations processes. Major existing legislative policies or actions that contribute to resilience are numerous and varied. Two foundational laws are the Stafford Act[4] and the Homeland Security Act of 2002[5]. These statutes provide most of the organizational and functional framework for mitigating, responding to, and recovering from natural disasters and acts of terrorism.

The most widely known law, and the most widely cited in the context of traumatic incidents, is the Stafford Act. The Stafford Act is intended:

> to provide an orderly and continuing means of assistance by the Federal Government to State and local governments in carrying out their responsibilities to alleviate the suffering and damage which result from such disasters. . . .[6]

Therefore, the Stafford Act is primarily a guide for responding to disaster incidents and does not refer explicitly to resilience.

Another piece of legislation, passed into law as The Disaster Mitigation Act of 2000 (P.L. 106-390), amended the Stafford Act:

> (1) to reduce the loss of life and property, human suffering, economic disruption, and disaster assistance costs resulting from natural disasters; and
> (2) to provide a source of predisaster hazard mitigation funding that will assist States and local governments (including Indian tribes) in implementing effective hazard

---

[4]The Robert T. Stafford Disaster Relief and Emergency Assistance Act (P.L. 100-707), signed into law on November 23, 1988; amended the Disaster Relief Act of 1974 (P.L. 93-288). The Disaster Mitigation Act of 2000 (P.L. 106-390) amended the Stafford Act and authorized a program for predisaster mitigation. The Stafford Act and its amendments constitute the statutory authority for most federal disaster response activities, especially as they pertain to FEMA and FEMA programs, https://www.fema.gov/library/viewRecord.do?fromSearch=fromsearch&id=3564.
[5]Homeland Security Act of 2002, P.L. 107-296, November 2002, http://www.dhs.gov/xabout/laws/law_regulation_rule_0011.shtm.
[6] https://www.fema.gov/library/viewRecord.do?fromSearch=fromsearch&id=3564.

> mitigation measures that are designed to ensure the continued functionality of critical services and facilities after a natural disaster.[7]

Thus, Congress recognized the need to prevent or minimize disasters, if possible, through hazard mitigation measures and provided funding mechanisms for that purpose, and that such measures need to be coordinated with, or performed by, state and local governments (FEMA, 2010).

The Homeland Security Act of 2002 was passed in the wake of the events of September 11, 2001, and created DHS, merging the structure and missions of 22 separate federal agencies. The Act sets forth the primary missions of the department, which are to

> (A) prevent terrorist attacks within the United States;
> (B) reduce the vulnerability of the United States to terrorism; and
> (C) minimize the damage, and assist in the recovery, from terrorist attacks that do occur within the United States.[8]

Although the new department's mission focuses on terrorism, DHS maintains responsibility for mitigating the effects of all kinds of disasters, including those from natural processes. Title V of the Act outlines those responsibilities

> "…..to reduce the loss of life and property and protect the Nation from all hazards by leading and supporting the Nation in a comprehensive, risk-based emergency management program—
> (A) of mitigation, by taking sustained actions to reduce or eliminate long-term risk to people and property from hazards and their effects;
> (B) of planning for building the emergency management profession to prepare effectively for, mitigate against, respond to, and recover from any hazard;
> (C) of response, by conducting emergency operations to save lives and property through positioning emergency equipment and supplies, through evacuating potential victims, through providing food, water, shelter, and medical care to those in need, and through restoring critical public services;
> (D) of recovery, by rebuilding communities so individuals, businesses, and governments can function on their own, return to normal life, and protect against future hazards; and

---

[7] http://www.disastersrus.org/fema/stafact.htm.

[8] http://www.dhs.gov/xabout/laws/law_regulation_rule_0011.shtm.

(E) of increased efficiencies, by coordinating efforts relating to mitigation, planning, response, and recovery."[9]

Although FEMA was placed within D, many of the traditional FEMA goals and activities continued to focus on natural hazards and an all-hazards approach to preparedness and response. The FEMA website states, "FEMA's mission is to support our citizens and first responders to ensure that as a nation we work together to build, sustain, and improve our capability to prepare for, protect against, respond to, recover from, and mitigate all hazards."[10] Thus, significant federal responsibility for some of the components of resilience building continues to lie within the mission of FEMA. However, the language of PPD-8 and the recommendations of the CRTF (see above) suggest that resources of DHS beyond FEMA are now expected to be brought to bear on the enhancement of national resilience.

Numerous policies to address specific components of community resilience have been introduced in Congress but have not been implemented; these bills nevertheless demonstrate cognizance of the need to strengthen specific aspects of resilience policy. For example, H.R. 2738, the Water Infrastructure Resiliency and Sustainability Act of 2011, has been introduced in the current Congress to address the supply and quality of water under conditions of climate change, a critical factor in the long-term resilience of communities.[11] Similarly, legislation has been introduced in the past that recognized the broader sweep of considerations that affect national resilience. For example, in 2003, H.R. 2370, the National Resilience Development Act, which did not become law, was intended to create an interagency task force on national resilience focused on "increasing the psychological resilience and mitigating distress reactions and maladaptive behaviors of the American public in preparation for and in response to a conventional, biological, chemical, or radiological attack on the United States."[12] Such efforts, though recognizing some of the most complex issues of resilience and worthy of consideration, do not address, in a comprehensive way, the myriad resilience issues simultaneously at work in communities.

Other laws contribute to resilience by addressing specific aspects of national hazards. For example, the National Earthquake Hazard Reduction Program (NEHRP)[13] provides for coordination among four federal agencies—FEMA, the National Institute of Standards and Technology, the National

---

[9] http://www.dhs.gov/xabout/laws/law_regulation_rule_0011.shtm.

[10] http://www.fema.gov/about/.

[11] Library of Congress, http://thomas.loc.gov/cgi-bin/thomas.

[12] National Institutes of Health, http://olpa.od.nih.gov/legislation/108/pendinglegislation/natresact.asp.

[13] NEHRP was created under the Earthquake Hazards Reduction Act of 1977, P.L. 95-124 (42 U.S.C. § 7701 et seq.), as amended by P.L. 101-614, P.L. 105-47, P.L. 106-503, and P.P. 108-360, http://www.nehrp.gov/about/PL108-360.htm.

Science Foundation, and USGS—to advance knowledge of earthquake causes and effects and to develop and promulgate measures to reduce their impacts at the community level, and the National Dam Safety Program, led by FEMA in coordination with other federal agencies, conducts research in dam safety, provides grants to 49 states to carry out state programs, and encourages individual and community responsibility for dam safety and related floodplain management.[14] These programs are examples of federal programs that are designed to understand the scientific underpinnings of natural hazards, to assess regional and local exposure to those hazards, and to communicate with the local communities to help them enhance their resilience to natural hazards. Arguably, increasing resilience at both the community and national levels is a central function of many of these federal programs.

STATE AND LOCAL AUTHORITIES AND POLICIES

A discussion of improved national resilience may lead to a discussion of federal policies, but many of the critical policies and actions required for improved national resilience must be enacted and implemented at the state and local levels. Federal policies and programs provide broad national direction across jurisdictions, but many aspects of community and state resilience lie completely outside the authority and purview of federal policy. As discussed in the previous chapter, the federal government has little or no jurisdiction over the local planning process, over zoning laws or building codes, or over numerous other critical aspects of local community resilience. The state and local authorities, the private sector, and individual citizens have key responsibilities and opportunities to improve resilience. This division of responsibility is not simply an oversight or an accident of governance. On the contrary, different responsibilities were assigned to the federal and state governments early in the nation's history, and the performance of specific functions by specific levels of governance arises from those principles.

At the local level, a number of jurisdictions and authorities may become involved in resilience planning, implementation, post-disaster recovery, and building, sometimes producing confusion or conflict about "who is in charge." During major events, the abilities and resources at the local level may be exhausted and aid is sought from state or federal government agencies and national organizations.

States derive their authority to govern the areas within their boundaries from the Tenth Amendment of the U.S. Constitution: "The powers not delegated to the United States by the Constitution, nor prohibited by it to the States, are reserved to the States respectively, or to the people."[15] States support the

---

[14] www.fema.gov/plan/prrevent/damfailure/ndsp.shtm.

[15] U.S. Constitution, http://www.archives.gov/exhibits/charters/bill_of_rights_transcript.html.

communities within their borders in a variety of ways, and most states, in turn, give local counties, cities, and municipalities limited authority through the so-called Dillon Rule (Virginia Natural Resources Leadership Institute, 2011), or broader authorities ("home rule") through their constitution or legislation.[16] Explicit coordination of disaster resilience planning and actions at the state level is not common across the United States, although a few states have begun to adopt specific approaches and establish offices to address the issue (Box 6.5). Home rule gives local communities broad authority to enact their own laws within the bounds of state and federal constitutions. The extent of local authority and how it is exercised is the subject of much debate and legal process, but most cities and towns have at least some authority to formulate community development plans and land-use plans, to institute zoning laws, to adopt and enforce building codes, and to pursue other measures to suit the resilience needs of their own community. Community leaders and elected officials, with the help and support of the public, local businesses and utilities, nongovernmental organizations, and perhaps with state and federal government assistance, will largely determine whether their community resilience increases, stays the same, or decreases.

BOX 6.5
Coordination of Resilience at the State Level

Following the Maryland Governor Martin O'Malley's service as chair of the U.S. Department of Homeland Security's Advisory Council's Community Resilience Task Force and experiences gained during Hurricane Irene, which cut a swath across the state, he established the Office of Resilience within the Maryland Emergency Management Agency (MEMA). The office was assigned the mission of bringing together the focused efforts of the state, the business sector, communities, nongovernmental agencies and other partners including faith-based groups and other volunteer organizations to deal with resilience development across the state.

The new office is developing a network for effective engagement in all areas of emergency management among the private- and public-sector entities, vulnerable populations, and relevant regional groups. They are carrying this out through aggressive outreach, education, planning and training efforts, and information sharing and needs identification. Much was learned from predisaster planned beneficial partnerships that were exercised following Hurricane Irene that were able to bring together the support of big box stores, supply chain facilitation in the food sector, and state efforts to limit impediments to interstate commerce by avoiding such things as hours-of-service limitations and road closures. The Executive Director of MEMA sees the new office as essential to

[16] http://definitions.uslegal.com/h/home-rule/.

fill a distinct need in dealing with disasters and one that will greatly improve resilience at all levels.
Sources: Richard Muth, Executive Director, MEMA, personal communication, March 26, 2012; Angela Bernstein, Director Office of Resilience, personal communication, April 3, 2012.

The role of the federal and state agencies is to assist local communities in these efforts. For example, FEMA uses tools such as its *Long-Term Community Recovery Planning Process: A Self-Help Guide* (FEMA, 2005) to help local communities plan their long-term recovery after a disaster, and NOAA assists coastal communities in becoming more aware of and more resilient to tsunamis.[17] Another approach, the Silver Jackets Program, was initiated by several federal agencies to reduce risk and increase resilience in a collaborative way with state and local agencies (Box 6.6). Many other federal programs provide similar guidance and assistance to local communities (see Table 6.1).

BOX 6.6
The Silver Jackets Program: Many Agencies—One Solution

The Silver Jackets[a] program is an innovative state-agency-centered effort initiated by the U.S. Army Corps of Engineers and the Federal Emergency Management Agency (FEMA) to bring together multiple state, federal, and local agencies (and where appropriate, tribes) to "learn from one another and apply their knowledge to reduce risk." It links the federal family of agencies with state and local counterparts as well as nongovernmental organizations (NGOs) to deal with challenging pre- and post-disaster issues. Programs are initiated at the state level and currently 29 states have such programs under way.

Its goals are to:

- Develop ways to "collaboratively address risk management issues, prioritize those issues, and implement solutions";
- Increase and improve risk communication through coordinated interagency efforts;
- Leverage available information and resources of all agencies such as FEMA's RiskMAP program and U.S. Army Corps of Engineers' (USACE) levee inventory and assessment initiative;
- Better coordinate hazard mitigation assistance by implementing in a collaborative manner those high-priority actions identified by state mitigation plans; and
- Identify gaps and conflicts among federal and state agency programs and provide recommendations for addressing these issues at both levels.

---

[17] National Tsunami Hazard Mitigation Program, http://nthmp.tsunami.gov/.

To deal with a need for flood mitigation, the Indiana Silver Jackets team has been supported by the U.S. Geological Survey (USGS) stream gauging program, a USACE planning assistance team, and the Department of Housing and Urban Development's Community Development Block Grant Program in assisting communities damaged by the 2008 Midwestern flood. Through this collaborative state–federal effort, the state will be able to improve flood warning systems and acquire LIDAR mapping for all 92 counties.

In Iowa, the Silver Jackets Team brings together the efforts of USACE's Rock Island and Omaha Districts, the National Weather Service, FEMA, USGS, the Natural Resources Conservation Service, and the Iowa Departments of Natural Resources, Emergency Management and Homeland Security, Agriculture and Land Stewardship, the Iowa Economic Development Authority, the Iowa Flood Center, the Iowa Utilities Board, and the Iowa Floodplain and Stormwater Management Association, an NGO. The team is currently dealing with issues in the Iowa-Cedar River watershed, including efforts to deal with the flood challenges of Cedar Rapids. When Cedar Rapids issues are under discussion, representatives from local agencies are included in the gatherings.

[a]Why *Silver Jackets*? Following a disaster, federal agencies frequently appear at the site wearing different colored jackets. The name *Silver Jackets* was proposed as way to reflect the collaborative efforts of all the agencies involved in pre- and post-disaster activities.
Source: www.nfrmp.us/state/about.cfm; Jerry Skalak, USACE -MVR, personal communication, March 29, 2012.

These principles and responsibilities that guide recovery also apply to developing community resilience more generally. For example, the recently released National Disaster Recovery Framework describes the roles and responsibilities for recovery, and the interactions of the different levels of government this way:

> Successful recovery requires informed and coordinated leadership throughout all levels of government, sectors of society and phases of the recovery process. It recognizes that local, State and Tribal governments have primary responsibility for the recovery of their communities and play the lead role in planning for and managing all aspects of community recovery. This is a basic, underlying principle that should not be overlooked by State, Federal and other disaster recovery managers. States act in support of their communities, evaluate their capabilities and provide a means of support for overwhelmed local governments. The Federal Government is a partner and facilitator in recovery, prepared to enlarge its role when the disaster impacts relate to areas where Federal jurisdiction is primary or affects national security. The Federal

> Government, while acknowledging the primary role of local, State and Tribal governments, is prepared to vigorously support local, State and Tribal governments in a large-scale disaster or catastrophic incident.[18]

However, many communities do not address, in a comprehensive manner, the numerous and complex issues that produce resilience until after a severe event occurs. The best time to develop resilience in a community is while the community is being planned and built or reconstructed after a disaster, and that is when the state and federal agencies may have somewhat limited roles. Therefore, it is critical that individuals and community leaders understand their roles and responsibilities relative to state and federal responsibilities, and that they consciously seek to improve the resilience of their community through their decisions and governing processes.

An example of building community resilience with specific local policies is through the implementation of resource planning policies by states and regional authorities that recognize threats from natural hazards also contribute to community resilience. For example, the State of Massachusetts recently adopted a climate change plan (Commonwealth of Massachusetts, 2011) to help avoid the consequences of anticipated changes resulting from climate change, and the San Francisco Bay Conservation and Development Commission (2011) issued a set of recommendations targeted at helping the San Francisco Bay area prepare for changes resulting from climate change and sea-level rise. Maryland has recognized the vulnerability of its coastal zones, particularly in light of the potential changes in sea level and climate, and has developed adaptation strategies for their coastal areas (Maryland Commission on Climate Change, 2008). Efforts such as these contribute to community and national resilience by identifying hazards and threats before a disaster occurs, allowing local administrations to adjust their development plans to protect their citizens.

UNINTENDED CONSEQUENCES: POLICIES AND PRACTICES THAT NEGATIVELY IMPACT RESILIENCE

Much of this chapter has focused on policies and programs that provide the framework for governance, responsibilities, and support of community resilience from the top down. But community resilience may also be affected by policies that are seemingly unrelated to resilience. Policies and practices

---

[18] http://www.fema.gov/national-disaster-recovery-framework, p. 9.

promulgated to address a wide variety of other national problems may have the unintended consequence of reducing resilience. Furthermore, in some cases, failure to enact a policy that would increase resilience results in a deterioration of resilience. In other words, the absence of a specific beneficial policy is, in itself, a policy. We present here a few examples of policies where unintended consequences have effectively reduced community resilience.

Agricultural policies provide one example of unintended consequences that reduce resilience. In this example, shifts in agricultural practice in the United States in response to farm policies designed to improve field drainage and productivity have unintentionally but significantly exacerbated flooding in the Midwest. Westward expansion of farming during the 19th century motivated farmers to improve the drainage in flat or low-lying farm fields to make them more productive. Improvement in field drainage was accomplished by the installation of drain tiles or perforated pipes just under the surface of the field to remove excess water. The effect of this accelerated drainage during the spring months of each year was to move water quickly from the fields to the streams and rivers, which exacerbated—and still exacerbates—flooding along many stream and rivers in the Midwest.

The contribution of field drainage to flooding was made even worse after the implementation of new agricultural policies following the Great Depression. As part of his suite of New Deal policies, President Franklin D. Roosevelt believed that true prosperity would not return to the nation until farming was prosperous. Roosevelt's Agricultural Adjustment Act of 1938 made federal price support mandatory for corn, cotton, and wheat and established permissible supports for many other crops and farm products.[19] The result of this policy was a fundamental shift in farming practice to row crops (mainly corn and soybeans) replacing traditional sod farming (perennial vegetation such as hay and densely sown small grains including oats, wheat, barley, triticale, and rye undersown with pasture grasses and legumes) as demonstrated for Iowa in Figure 6.1 (Jackson, 2002; see also Mutel, 2010).

---

[19] Agricultural Adjustment Act, P.L. 75-430, United States Code, Title 7, Chapter 35, http://frwebgate.access.gpo.gov/cgi-bin/usc.cgi?ACTION=BROWSE&TITLE=7USCC35&PDFS=YES.

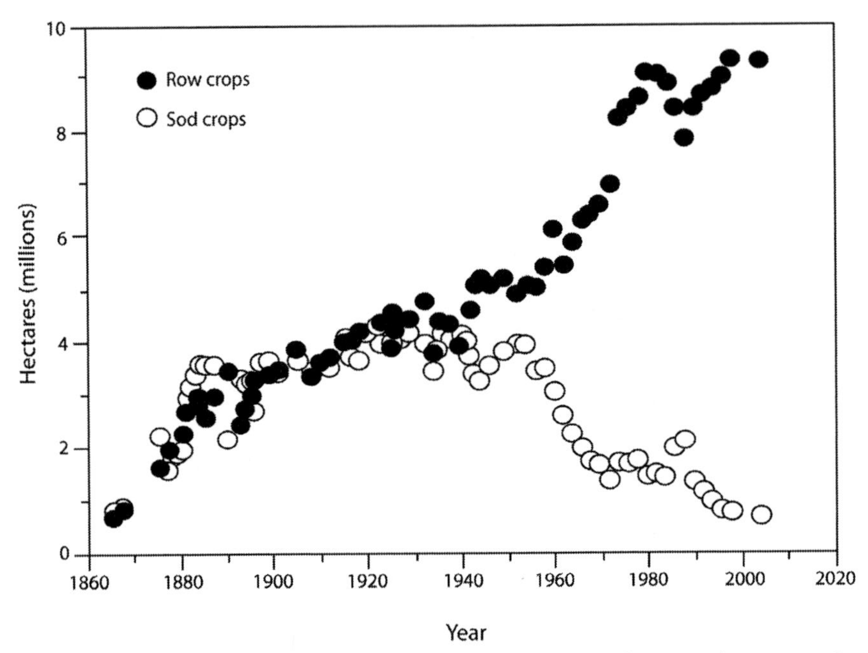

FIGURE 6.1 Shift in farming practice in Iowa to row crops from earlier focus on sod crops around 1938 as a result of the Agricultural Adjustment Act. Source: Adapted from Jackson (2002).

For more than 60 years (1870 to the 1930s) Iowa farmers had maintained about 50 percent sod crop, but with passage of the Agricultural Adjustment Act of 1938 row crops began to dominate, with dramatic implications for flood resilience (Jackson, 2002). The traditional sod crops had dense root masses that absorbed rainfall without runoff and released it back to the atmosphere via transpiration and through underground flow into both shallow and deep aquifers (Jackson and Keeney, 2010). Because the crops were perennial, after harvest the root mass remained and was not tilled up, thus retaining and improving top soil. Knox (2006) describes the agricultural conversion of prairie and forest in the upper Mississippi Basin as the most important environmental change that influenced fluvial (river and stream) activity in this region in the past 10,000 years.

Even without impacts of climate change, farm practice (responding in part to policy) has significantly increased the flood potential in the Midwest. The overall effect of facilitating the drainage of millions of acres of farm fields through underground drains, combined with the shift from sod crops to row crops and the encroachment of many communities into the floodplain, was to reduce the resilience of cities and towns along Midwestern rivers by increasing the likelihood and intensity of flooding. To address this problem, Jackson and Keeney (2010) summarize a variety of proposed novel mitigation strategies including crop rotation, strip-cropping practice, crop mixing, as well as setting aside small percentages of row-crop land for perennial "buffer strips" along streams. This example, like many others, contains many variables and many forces, and cannot be distilled into a simple choice between flooding and soggy fields or subsidies that encourage unsustainable farming practices, but it serves to demonstrate that unintended consequences of well-intentioned national agricultural policies may ultimately reduce local resilience.

Forest management policy provides a second example of unintended consequences of policies or practices. A century of aggressive suppression of wildland fires combined with recent broad and extended periods of drought, have substantially altered many of the nation's forests and have resulted in devastating wildfires at the wildland–urban interface in many locations across the United States. These fires are difficult to control, threaten adjacent urban areas, and are expensive to fight (Cohen, 2008). Corrective policies that emphasize fuel management are often underfunded or infeasible. In their review, USDA ecologists Donovan and Brown (2007) recommend a different approach to wildfire management that focuses on encouraging managers to balance short-term wildfire damages against the long-term consequences of fire exclusion. The approach deemphasizes fire suppression. Recent changes in the management of wildland fires recognize the effects of past policies on forested communities and these new policies increase the resilience of those communities and accommodate the sustainability of ecosystems (National Wildfire Coordinating Group, 2009).

Likewise, government policies for coastal zone management have traditionally been intended to balance economic development along the coasts with preservation of coastal habitat and environment while recognizing the risks of development along the coast.[20] Now more than 50 percent of the U.S. population lives within 50 miles of a coastline and this proportion is expected to increase in the future.[21] Economic development, including residential, commercial, recreational, and industrial development in the coastal zone has greatly increased the exposure to storm surge, coastal erosion, and sea-level rise. Federal policy for coastal zones has been to encourage and support coastal states in the proper development and management of their coastal areas, but some states have placed short-term economic development above long-term safety and community resilience.

Perhaps the classic example of unintended consequences of well-intentioned historical policies is the effects of Mississippi River flood management on the City of New Orleans and the Mississippi River delta communities. This series of historical decisions and engineering efforts has been thoroughly documented in several publications (Coastal Protection and Restoration Authority of Louisiana, 2012). Many decades of efforts to levee and channel the Mississippi River to reduce flooding and facilitate navigation along

---

[20] Coastal Zone Management Act of 1972, as amended through P.L. 109-58 and the Energy Policy Act of 2005, http://coastalmanagement.noaa.gov/czm/czm_act.html.

[21] NOAA, http://oceanservice.noaa.gov/facts/population.html.

the course of the river as well as the construction of large dams on the main stem of the Missouri River combined with construction of channels for transportation of oil and gas exploration have starved the Mississippi River delta of sediment and have resulted in increased vulnerability to tropical storms and hurricanes in the Mississippi delta region. The normal natural processes of sedimentation and delta growth were halted and the subsidence of the delta edifice was not counteracted by the deposition of new sediments across the delta. The result is a subsiding and shrinking delta with reduced capacity to mitigate storm surge. These effects have severely degraded the resilience of the delta and the human settlements in the region, including New Orleans. These historic policies have made the entire Mississippi delta region less resilient.

In addition to unintended consequences of individual policies, the lack of communication and coordination among federal agencies may have real consequences for communities or victims of a disaster. Sometimes an individual policy may be beneficial, but when multiple federal agencies independently apply mutually unknown policies to the same geographic area or structure, those policies may be contradictory and may inhibit recovery or slow the enhancement of resilience. For example, if one agency bases the distribution of funds on the value of a property on a floodplain at the same time that a policy of a different agency is changing the value of that property through acquisition or demolition, the property owner may be caught in a quandary and may be excluded from a funding mechanism through no action or fault of his or her own. The application of federal policies either before or after disasters needs to be informed by the goals of the community and by the knowledge of other policies that are being applied by other agencies. This coordinated application of policies will only be achieved if communication and coordination among federal agencies is achieved, and if agencies are aware of the needs and priorities of the affected community or individual.

An unintended consequence of certain security policies adopted after the September 11, 2001 World Trade Center attack is the difficulty of some local governments and the private sector in gaining access to certain information necessary to secure privately owned infrastructure against various hazards and to develop plans to deal with emergency events. A report on National Dam Safety to FEMA by the University of Maryland identified the restrictions placed on release of information on dam integrity and potential downstream inundation as significant impediments to disaster planning and preparedness (Water Policy Collaborative, 2011). A 2012 Report by the National Research Council on dam and levee safety and community resilience similarly concluded that

> Those subject to the direct or indirect impacts of dam or levee failure are also those with the opportunity to reduce the consequences of failure through physical and social changes in the community, community growth planning, safe housing construction, financial planning (including bonds and

> insurance), and development of the capacity to adapt to change. (NRC, 2012, p. 107)

As pointed out by Flynn and Burke (2011), investment and operational decisions by corporations that own critical infrastructure may be made without full security awareness because information that has been classified by the Department of Homeland Security is sometimes not available to the corporate executives making the decisions. Because an increase in community resilience requires coordination and cooperation among all key players within the community, including the private-sector owners of infrastructure, it is vitally important that communities be aware of prescribed rules and methods of sharing restricted information in a secure way among all partners, including the vital private-sector partners, as detailed in Executive Orders 12829,[22] 12958,[23] and 13292.[24] Some types of data may be sensitive, but giving local partners the opportunity to work with state and federal stakeholders on equal footing is important to build long-term resilience.

Finally, even some policies that seem unrelated to community or national resilience may unintentionally and negatively affect resilience. A recent example of this is the Budget Control Act of 2011. The President signed the Budget Control Act of 2011 into law (P.L. 112-25) on August 2, 2011. The purpose of that legislation is primarily to increase the U.S. debt limit, establish caps on the annual appropriations process over the next 10 years, and to create a Joint Select Committee on Deficit Reduction that is instructed to develop a bill to reduce the federal deficit over the 10-year period. One provision of this new law that affects U.S. national resilience is an amendment to Section 251 of the Balanced Budget and Emergency Deficit Control Act of 1985. That amendment provides for disaster relief appropriations each fiscal year based on "the average funding provided for disaster relief over the previous 10 years, excluding the highest and lowest years." In this bill, "the term 'disaster relief' means activities carried out pursuant to a determination under section 102(2) of the Robert T. Stafford Disaster Relief and Emergency Assistance Act (42 U.S.C. 5122(2))." As discussed elsewhere in this report, developing national resilience encompasses more elements than disaster recovery alone. Building a resilient community requires thoughtful and strategic long-term investments in multiple aspects of the physical and social fabric of communities that contribute to resilience. Of course, disaster recovery is an integral part of that process because the ability of communities to recover after a disaster, and the way that they recover, is closely tied to becoming more resilient to subsequent trauma. Therefore, the federal commitment to assist communities in a timely fashion is central to the long-term resilience of communities. When a community's

---

[22] http://www.archives.gov/isoo/policy-documents/eo-12829.html.

[23] http://www.fas.org/sgp/clinton/eo12958.html.

[24] http://www.archives.gov/isoo/policy-documents/eo-12958-amendment.html.

capacity to respond to a disaster is overwhelmed, its very survival depends on how recovery is conducted. If resources are delayed or curtailed during the critical recovery phase of a disaster, it is possible that states, local communities, businesses, and neighborhoods may be unable to rebuild in a resilient way (or not at all) and even greater costs will result over the long-term.

RESILIENCE POLICY GAPS AND NEEDS

Recognizing that community resilience is advanced by a variety of policies at the federal, state, and local levels, combined with corporate policies and practices, it is important to ask what policies might improve resilience. What policies are absent and badly needed? What new policies should be adopted at each level of government to continue the improvement in the resilience of U.S. communities? Federal policies to strengthen the resilience of communities may be broad or narrow, short term or long term. Because resilience grows over the long term through the application of principles and policies that guide local decisions, the most fruitful policies will be those that acknowledge the broad, long-term needs of communities. Although identification of specific resilience policy gaps is essential to advancing the nation's resilience, an a la carte approach to resilience policy, in the absence of an overall national strategy, may result in contradictory policies or gaps. Strong communication and coordination among agencies and stakeholders will help ensure effective actions.

The nature of resilience requires some flexibility and adaptability because the patterns of risk, development, and culture vary so widely among communities (see also Chapters 3 and 5). Consideration of this need for flexibility is important for policymakers pursuing mechanisms to enhance the resilience of communities. The fluid and progressive nature of seeking a resilient community does not lend itself to laws or policies mandating resilience as a perfect condition of a community. Any federal, state, or local policies that attempt to mandate resilience would imply that resilience is a perfectly definable condition, which it is not. Community resilience is highly desirable, but broadly complex, and would be extremely difficult to codify in a single comprehensive law.

Rather, governments at all levels have to formulate their own visions of resilience and take the steps in all of their processes to advance resilience through all of its components, forms, and functions, and seek to infuse the principles of resilience into all routine functions of the government. Some ways in which this might be done is the topic of the next chapter.

Currently, gaps in policies and programs exist among federal agencies for all parts of the resilience process—including disaster preparedness, response, recovery, mitigation, and adaptation, as well as research, planning, and community assistance. Although some of these gaps are the result of the legislative authorization within which agencies are directed to operate, the roles and responsibilities for building resilience are not effectively coordinated by the federal government, either through a single agency or authority or through a unified vision about how these roles and responsibilities for promoting resilience could be organized. The roles and responsibilities in the federal government for long-term recovery and improvement of resilience constitute a particularly significant policy gap despite some recent legislation and initiatives. Implementation of PPD-8 should help address this gap. At the state and local levels, many jurisdictions have made excellent progress in taking both a long and broad view of community resilience, and these communities can be used as models. However, many local communities find themselves torn among competing priorities, and the advancement of long-term community resilience is often undermined by the need or desire to address an urgent condition or opportunity in the community. Clearly, policies and processes to improve national resilience at all levels of government will improve as the benefits of resilience are realized and the efforts to improve resilience are integrated across jurisdictions.

SUMMARY, FINDINGS, AND RECOMMENDATION

Leaders at the local, state, and federal level are increasingly aware of community resilience and how it might be advanced through a variety of decisions and processes. Although many of those critical decisions and processes to improve resilience occur at the state and local levels, the federal government plays a central role in providing guidance for policy and program development to assist local communities in their pursuit of greater resilience. Development of new policies can be informed by an awareness of resilience, how it can be promoted through decisions and processes, and how resilience can be unintentionally eroded through poorly informed decisions.

Three significant findings from the assessment of the policy landscape of resilience are:

(1) The development of appropriate policies, creation of optimal governance structures, and informed and coordinated management at all levels of government are crucial to improving community resilience. Community resilience will grow as the knowledge, experience, and understanding of these roles and responsibilities grow among decision makers at all levels of government.
(2) Currently a multitude of activities, programs, and policies exist at local, state, and federal levels to address some part of resilience for the nation. Several of the critical processes, such as land-use planning and building code enforcement, are the responsibility of local groups or governments. The federal policy role is primarily to ensure that resilience policies are nationally consistent and to provide information and best practices for

development of appropriate policies at all levels. Consideration of potential unintended consequences of a new policy with respect to disaster resilience is also important.

(3) The nation does not currently have an overall vision or coordinating strategy for resilience. Recent work on homeland security and disaster reduction are good beginnings, but the current suite of policies, practices, and decisions affecting resilience are conducted on an ad hoc basis with little formal communication, coordination, or collaboration. In fact, some policies, decisions, and practices actually erode resilience. Implementation of PPD-8 will address some of these consistency and coordination issues.

Recommendation: **All federal agencies should ensure they are promoting and coordinating national resilience in their programs and policies. A resilience policy review and self-assessment within agencies and strong communication among agencies are keys to achieving this kind of coordination.**

Such an assessment should reveal how each agency's mission contributes to the resilience of the nation, and how its programs provide knowledge or guidance to state and local officials for advancing resilience. Finally, each federal agency should evaluate its interactions with state and local governments and with the public to evaluate the extent to which its resilience work is made available to those who need it.

REFERENCES

Coastal Protection and Restoration Authority of Louisiana. 2012. Louisiana's Comprehensive Master Plan for a Sustainable Coast. Available at http://www.lacpra.org/assets/docs/2012%20Master%20Plan/Final%20Plan/2012%20Coastal%20Master%20Plan.pdf.

Cohen, J. 2008. The wildland-urban interface fire problem. *Forest History Today* (Fall):20-26. Available at: http://www.foresthistory.org/publications/FHT/FHTFall2008/Cohen.pdf.

Commonwealth of Massachusetts. 2011. Climate Change Adaptation Report. Executive Office of Energy and Environmental Affairs and the Adaptation Advisory Committee Massachusetts. Available at http://www.mass.gov/?pageID=eoeeaterminal&L=3&L0=Home&L1=Air%2C+Water+%26+Climate+Change&L2=Climate+Change&sid=Eoeea&b=terminalcontent&f=eea_energy_cca-report&csid=Eoeea.

CRTF (Community Resilience Task Force, Homeland Security Advisory Council). 2011. Community Resilience Task Force Recommendations. Available at http://www.dhs.gov/xlibrary/assets/hsac-community-resilience-task-force-recommendations-072011.pdf.

Donovan, G. H., and T. C. Brown. 2007. Be careful what you wish for: The legacy of Smokey Bear. *Frontiers in Ecology and the Environment* 5(2):73-79. Available at http://gis.fs.fed.us/rm/value/docs/legacy_smokey_bear.pdf.

FEMA (Federal Emergency Management Agency). 2005. Long-Term Community Recovery Planning Process: A Self-Help Guide. Available at: http://www.fema.gov/library/viewRecord.do?id=2151. Accessed March 12, 2012.

FEMA. 2010. Hazard Mitigation Assistance Unified Guidance. Available at http://www.fema.gov/library/viewRecord.do?id=4225.

Flynn, S., and S. Burke. 2011. Brittle infrastructure, community resilience, and national security. *TR News* 275(July-August):4-7. Available at http://cnponline.org/index.php?ht=a/GetDocumentAction/i/34968.

Jackson, L. L. 2002. Restoring prairie processes to farmlands. In *The Farm as Natural Habitat: Reconnecting Food Systems with Ecosystems*, Washington, DC: Island Press, pp. 137-154.

Jackson, L., and D. Keeney. 2010. Perennial farming systems that resist flooding. In *A Watershed Year: Anatomy of the Iowa Floods of 2008,* C. F. Mutel, ed. Iowa City: University of Iowa Press. Available at http://www.uiowapress.org/books/2010-spring/mutel-water.htm.

Knox, J. C. 2006. Floodplain sedimentation in the upper Mississippi Valley: Natural versus human accelerated. *Geomorphology* 79:286-310.

Longstaff P. H., N. J. Armstrong, K. Perrin, W. M. Parker, and M. A. Hidek. 2010. Building resilient communities: A preliminary framework for assessment. *Homeland Security Affairs* 6(3). Available at http://www.hsaj.org/?fullarticle=6.3.6.

Maryland Commission on Climate Change. 2008. Climate Action Plan, Interim Report to the Governor and the Maryland General Assembly, January 14. Available at http://www.mdclimatechange.us/ewebeditpro/items/O40F14798.pdf.

Mutel, C. F., ed. 2010. *A Watershed Year: Anatomy of the Iowa Floods of 2008*. Iowa City: University of Iowa Press. Available at http://www.uiowapress.org/books/2010-spring/mutel-water.htm.

National Wildfire Coordinating Group. 2009. Review and Update of the 1995 Federal Wildland Fire Management Policy. Available at http://www.nwcg.gov/branches/ppm/fpc/archives/fire_policy/docs/exsum.pdf.

NOAA (National Oceanic and Atmospheric Administration). 2012. Chart the Future: NOAA's Next Generation Strategic Plan. Available at http://www.ppi.noaa.gov/wp-content/uploads/NOAA_NGSP.pdf.

NRC (National Research Council). 2010. *Limiting the Magnitude of Climate Change.* Washington, DC: The National Academies Press.

NRC. 2012. *Dam and Levee Safety and Community Resilience: A Vision for Future Practice.* Washington, DC: The National Academies Press.

San Francisco Bay Conservation and Development Commission. 2011. Resolution No. 11-08: Adoption of Bay Plan Amendment No. 1-08. Available at http://www.bcdc.ca.gov/proposed_bay_plan/bp_amend_1-08.shtml.

White House and DHS (White House and the Department of Homeland Security). 2011. Presidential Policy Directive-8. Available at http://www.dhs.gov/xlibrary/assets/presidential-policy-directive-8-national-preparedness.pdf.

USGS (U.S. Geological Survey). 2009. USGS Science: Addressing Our Nation's Challenges, General Information Product 93. Available at http://pubs.usgs.gov/gip/93/.

Virginia Natural Resources Leadership Institute. 2011. The Dillon Rule: Legal Framework for Decision-Making. Available at http://www.virginia.edu/ien/vnrli/docs/briefs/DillonRule_2011.pdf.

Water Policy Collaborative, University of Maryland,. 2011. Review and Evaluation of the National Dam Safety Program. Available at http://www.fema.gov/library/viewRecord.do?fromSearch=fromsearch&id=5794.

7

Putting the Pieces Together: Linking Communities and Governance to Guide National Resilience

National resilience rests on a foundation of choice—and begins with communities and how cities, towns, and the landscapes on which they exist are planned, designed, constructed, and maintained. This foundation is the physical resilience (Poland, 2012). Resilience then grows out of the health, security, and well-being of our people, which is a combined effort and responsibility of the people in the communities and the governing bodies—at all levels—that develop and implement resilience-building policies. This human component has been called an "engine" that can drive the physical foundation forward to increase resilience (Poland, 2012; see also Chapters 5 and 6). Because resilience cannot be accomplished by simply adding a cosmetic layer of policy or practice to a vulnerable community, long-term shifts in physical approaches (new technologies, methods, materials, and infrastructure systems) and social practices and initiatives (the people, management processes, institutional arrangements, and legislation) are needed to advance community resilience.

Communities and the governance network of which they are a part are complex and dynamic systems. Resilience to disasters rests on the premise that these multiple systems are robust, and requires that the system components work in concert and in such a way that the interdependencies provide strength during a disaster event. Experience in the disaster management community suggests that linked bottom-up–top-down networks are important for managing risk and increasing resilience (IPCC, 2012). Institutionally driven or top-down arrangements may in fact constrain or otherwise impede local actions if links or networks are not made to community-based or bottom-up approaches (Cutter et al., 2012). The dynamic nature of communities lends itself to comparisons to organisms, such as the human body metaphor used in Chapter 1. This suggests a holistic rather than piecemeal approach toward enhancing the nation's resilience. Because of the cost and commitment needed to increase resilience, two potential paths were outlined in Chapter 1 to consider in addressing the nation's approach to resilience. One path was that of investment in a long-term strategy of increasing the nation's resilience through concerted collaboration and action on the part of governing bodies and the communities they serve. The other path was one of maintenance—where current policies and approaches are continued without a long-term view, and disasters are addressed in a reactive way as they arise. Although the first path toward increased resilience may be more expensive initially, other chapters have presented evidence that longer-term savings can result from such an approach.

Although improvements in the nation's physical resilience (Chapters 2-4) are needed, the committee sees the interactions among and actions by the communities and governing bodies described in Chapters 5 and 6 as keys (the "engines") to move resilience forward. The committee has observed and documented numerous cases of individual success (at the level of a community, a government agency, a city) in taking steps toward increasing resilience; however, Chapters 5 and 6 make evident the fact that the collective, national resilience "engine" is not running optimally to make significant advances in resilience across the country.

Table 7.1 attempts to capture visually some of the key interactions within the nation's resilience system by identifying specific kinds of policies that can increase resilience and the roles and responsibilities of those in government, the private sector, and communities for acting on these policies. The purpose of the table is two-fold: (1) it attempts to visualize in a relatively simple way the complex interactions and dependencies in the community resilience system—one that combines bottom-up and top-down approaches; and (2) it attempts to show policy areas where the nation is currently making some progress toward communicating or implementing a type of resilience-building policy. By framing some of the collective responsibilities and identifying some of the gaps in the collaborative resilience network, the committee aims to help direct future discussion among these various stakeholders toward those areas of resilience building that may need most immediate attention.

TABLE 7.1 Overview of some key resilience policies and the roles and responsibilities related to developing and maintaining community resilience.

Policy Goal	Policy Area	Resilience Policy Result	Stakeholders					
			FED	STA	REG	LOC	PRIV	COMM
Strong governance, all levels								
	Governance	Agency roles and responsibilities clearly defined	√++	√++	√++	√++	O	O
		Jurisdictional cooperation	√++	√++	√++	√++	O	O
		Preparedness, planning, training	√+	√++	√+	√++	√	O
		Response	√++	√++	√√	√++	√	O
Resilience-aware leaders and citizens								

Policy Goal	Policy Area	Resilience Policy Result	Stakeholders					
			FED	STA	REG	LOC	PRIV	COMM
	Education/ Community preparedness	Resilience awareness	√++	√++	√	√++	√√	√++
		Community organization, planning, preparedness	√√	√√	√+	√++	√+	√+
		Preparation for response, recovery	√++	√++	√+	√+	√√	√
Stable and resilient economy								
	Economics/ finance	• Aid to state and local government • Preparation, building resilience • Response	√++	O	O	O	√+	O
		Recovery costs	√+	√++	O	√++	√+	O

Policy Goal	Policy Area	Resilience Policy Result	Stakeholders					
			FED	STA	REG	LOC	PRIV	COMM
		Insurance	√√	√+	√	√	√++	O
		Poverty reduction	√++	√++	√++	√++	√++	√√
Understanding threats and processes								
	Science & Technology/ R&D	Understanding/ forecasting disasters	√++	√√	O	O	√++	O
		Detection/ monitoring	√++	√++	O	O	O	O
		Geospatial information	√++	√+	√√	√√	√++	O
	Energy/ Climate change	Control carbon emissions	√++	√++	√	√++	√++	√++
	Coastal management	Manage coastal development	√+	√++	√	√++	√+	√+

Policy Goal	Policy Area	Resilience Policy Result	Stakeholders					
			FED	STA	REG	LOC	PRIV	COMM
Resilient infrastructure and landscape								
	Planning and assessment tools	National resilience dataset	√++	√√	√√	√√	√+	O
	Land-use planning	Manage development of vulnerable lands	√++	√++	√++	√++	√++	√++
	Zoning	Establish rules for development	√√	√√	√√	√++	O	O
		Enforce rules for development	√√	√√	√+	√++	O	O
	Codes and standards	Develop rules for buildings	√++	√+	√+	√++	√+	√+
		Enforce rules for buildings	√+	√	√	√++	O	O

Policy Goal	Policy Area	Resilience Policy Result	Stakeholders					
			FED	STA	REG	LOC	PRIV	COMM
	Critical infrastructure	National standards: Prioritize protection and hardening, interdependency analysis	√++	√+	√	√++	√+	O
		Emergency shelters	√√	√+	√	√+	√+	O
Strong citizen protection								
	Public health, human services	Medical records availability	√+	√+	√	√+	√+	O
		Access to health care including financial coverage	√++	√√	√	√√	√++	O
		Locator systems for reuniting people	√++	√+	√	√+	√++	O
Strong communications								

Policy Goal	Policy Area	Resilience Policy Result	Stakeholders					
			FED	STA	REG	LOC	PRIV	COMM
	Communica-tions	Media	√++	√+	√	√+	√++	O
		Social media	√	√	√	√+	√++	√++
		Resilience awareness	√++	√++	√++	√++	√++	√++
		Warning systems	√++	√+	√+	√+	√√	O

Notes: Roles are identified for key actors in federal, state, regional, and local government, the private sector (for-profit), and communities (non-profit and individuals). A check symbol "√" in the box indicates that the given actor has a responsibility in that area. An "O" in the box indicates an actor is a recipient or target for that policy. The additional symbols associated with the checks indicate the level of coordinating responsibility of that actor relative to the other actors, on a four-level scale: (1) √++, primary coordinating responsibility (in cases where multiple actors are designated primary responsibility, each actor is assumed to have responsibility within its purview; (2) √+, significant responsibility but not primary; (3) √√, minor responsibility; (4) √, participatory, rather than primary, coordinating responsibility. Responsibility designations do not give any indication of how well or successfully the responsibility is carried out, or if other stakeholders might best be given more or less responsibility. That type of analysis may be useful for the various stakeholders to conduct as they evaluate their own policy goals and roles of partners in increasing resilience. Fed = Federal; Sta = State; Reg = Regional; Loc = Local; Priv = Private; Comm = Community.

The committee's goal in this study was not to provide a set of complete solutions toward increasing the nation's resilience. Rather, the study places resilience in the context of practical physical and human elements that are critical to the nation in attempting to advance disaster resilience. Advancing resilience is a long-term process, but can be coordinated around visible, short-term tasks that allow individuals and organizations to mark their progress toward becoming resilient. The practical recommendations in Chapter 8 attempt to identify some of these long- and short-term approaches. As a necessary first step to strengthen the nation's resilience and provide the leadership to establish a national "culture of resilience," a full and clear commitment to disaster resilience by the federal government is essential.

Recommendation: **Federal government agencies should incorporate national resilience as a guiding principle to inform the mission and actions of the federal government and the programs it supports at all levels.**

The breadth and potential fragmentation of federal activities related to disaster resilience require a clear vision for national resilience, including federal roles and responsibilities, and a comprehensive strategy for advancing resilience of communities, institutions, and sectors. The broad framework and principles for this vision and strategy should derive from the Executive Branch. PPD-8 and the subsequent recommendations of the Community Resilience Task Force (Chapter 6) provide a strong beginning for such a federal vision. Such a vision for the nation includes participation and input from the local and private-sector stakeholders, and can serve as a template for similar visions and strategies developed by states, regions, cities, and neighborhoods for their respective communities and tailored by them for their needs and priorities.

The acceptance of a shared vision and a shared responsibility at the federal level is a critical step in achieving national resilience. Development and implementation of the vision can be achieved in part by the federal agencies through clear definition of their individual and collective roles and responsibilities and their roles in promoting resilience among state and local governments, the business community, nongovernmental organizations (NGOs) and nonprofits, and local communities. Clear definition of federal roles for resilience would also allow communities to understand their own roles and responsibilities for promoting resilience and would provide the basis for dialogue with federal agencies to address all phases of the resilience process and to close gaps that presently exist in the process. Some potential steps to implement a national resilience vision and strategy are outlined below.

STEPS FOR IMPLEMENTATION

All federal agencies are responsible for increasing resilience and for developing the national resilience vision, although different agencies will take the lead for various aspects of resilience.

In PPD-8, the President directs the Secretary of Homeland Security to "coordinate a comprehensive campaign to build and sustain national preparedness, including public outreach and community-based and private-sector programs to enhance national resilience, the provision of Federal financial assistance, preparedness efforts by the Federal Government, and national research and development efforts" (White House and DHS, 2011). Through PPD-8, the Department of Homeland Security (DHS) is directed to assume a broad coordination and leadership role that brings national resilience into focus at the federal level, and provides clear and coordinated collaboration with state and local government, the private sector, and individuals. The coordination of public outreach, federal financial assistance, preparedness efforts by other federal agencies, and resilience-related research and development efforts across the government is a necessary responsibility for DHS and all relevant federal agencies to pursue aggressively. A group of federal agencies convened under the presidential authority of PPD-8 should address the following **short-term tasks** (in the 1- to 2-year time frame) to incorporate resilience as an organizing principle in federal agency missions and actions:

(1) Develop a national vision of resilience:

- Develop, with participation of state, local, and private-sector stakeholders, a vision of national resilience to serve as a foundation for longer-term discussion of a national vision to be shared with communities (at state, regional, and local levels).
- Define, within each federal agency, resilience-related roles, responsibilities, and key ongoing activities, especially as related to existing efforts related to homeland security and disaster reduction;

(2) Develop a communications strategy to promote resilience among federal agencies, state and local government, and other stakeholders, including NGOs such as hospitals, religious communities, aid agencies, schools, and universities. Communications could be aided by provision of a real or virtual forum for all community stakeholders to share knowledge, experience, and needs among those focusing on national resilience.

(3) Develop and facilitate an effective coordination, collaboration, and accountability process for resilience planning and implementation among federal agencies. The current efforts in homeland security and disaster reduction support such coordination among federal agencies, but a focus on long-term planning, policy impacts, and gathering input from state and local

authorities and groups would enhance coordination, collaboration, and accountability.

(4) Conduct an analysis of federal, state, and local funding for disaster preparedness and response, including all natural hazards and critical infrastructure investments, and develop a cost-effective strategy for short- and long-term investments in the components of resilience.

(5) Identify achievable **long-term tasks** (in the 3- to 10-year time frame) to fully implement the shared national resilience vision that include, for example,

- Establishing a process for dialogue, planning, and coordination among local, state, and national government leaders and agency heads to develop a long-term national resilience implementation strategy. This process could include:
  - protocols and processes for data collection and data management;
  - coordinating funding streams to local communities for resilience enhancement for case management during and following disasters, for preparedness, response, and short- and long-term recovery; and
  - developing appropriate metrics and a process for measuring progress in advancing national resilience.
- Developing short-term incentives and guideposts for achieving these long-term goals. Such incentives can address the tendency for decision makers to focus on short-term horizons.
- Developing a consistent and coordinated communication and outreach strategy around the national vision for resilience for the general public.
- Developing a long-term investment strategy for federal funding of resilience priorities within the context of existing funding of disaster preparedness and response.
- Conducting periodic review and assessment of agency activities to assess progress in the implementation goals and strategies of the national resilience vision.

REFERENCES

Cutter, S., B. Osman-Elasha, J. Campbell, S.-M. Cheong, S. McCormick, R. Pulwarty, S. Supratid, and G. Ziervogel. 2012. Managing the risks from climate extremes at the local level. In *Managing the Risks of Extreme Events and Disasters to Advance Climate Change Adaptation. A Special Report of Working Groups I and II of the Intergovernmental Panel on Climate Change (IPCC)*, C. B. Field, V. Barros, T. F. Stocker, D. Qin, D. J. Dokken, K. L. Ebi, M. D. Mastrandrea, K. J. Mach, G.-K. Plattner, S. K. Allen, M. Tignor, and P. M. Midgley, eds. , Cambridge, UK, and New York: Cambridge University Press, pp. 291-338.

IPCC (Intergovernmental Panel on Climate Chante). 2012. Summary for policymakers. In *Managing the Risks of Extreme Events and Disasters to Advance Climate Change Adaptation. A Special Report of Working Groups I and II of the Intergovernmental Panel on Climate Change (IPCC)*, C. B. Field, V. Barros, T. F. Stocker, D. Qin, D. J. Dokken, K. L. Ebi, M. D. Mastrandrea, K. J. Mach, G-K. Plattner, S. K. Allen, M. Tignor, and P. M. Midgley, eds. , Cambridge, UK, and New York: Cambridge University Press, pp. 3-21.

Poland, C. 2012. Preparing to Recover: Creating a Disaster-Resilient Community. Presentation to the Committee, February 3. Available at http://www.cvent.com/events/earthquakes-mean-business-2012/agenda-dae7fa1644c84f93ada7af6df4f3e5a7.aspx.

White House and DHS (White House and Department of Homeland Security). 2011. Presidential Policy Directive-8. Available at http://www.dhs.gov/xlibrary/assets/presidential-policy-directive-8-national-preparedness.pdf.

8

Building a More Resilient Nation: The Path Forward

Natural and human-induced disasters carry with them the potential for injuries and death, displacement of people, loss of homes and land, disruptions in transportation, business interruption, job losses, and greater demands on federal, state, and local resources. Against the backdrop of the nation's aging infrastructure, inconsistent adoption and enforcement of building codes, and health and economic disparities, the future impacts of global population growth and movement, complex and interdependent global commerce and economic systems, and changing climate demand greater resilience to disasters to help decrease disaster-related losses and to increase the nation's physical, social, cultural, economic, and environmental health.

This chapter draws together the six recommendations made in earlier chapters and provides suggestions as to how these recommendations might be implemented. The committee has indicated that the necessary first step to increased resilience is to establish a national "culture of resilience" which includes a full and clear commitment to disaster resilience by the federal government.

***Recommendation 1*: Federal agencies should incorporate national resilience as a guiding principle to inform the mission and actions of the federal government and the programs it supports at all levels.**

This recommendation embodies an approach that includes development of a national vision and a national strategy toward a more resilient nation, and a set of short- and long-term implementation steps to achieve this goal including:

(a) Development of the resilience vision;
(b) Development of communication strategies for promoting resilience among federal, state, and local governments, communities, and the private sector;
(c) Analysis of appropriate investment strategies for increasing resilience;

(d) Establishment of processes for interagency coordination for data and resilience metrics;
(e) Establishment of incentives for increasing resilience; and
(f) Conducting periodic reviews of federal agency progress toward increasing resilience (see Chapter 7 for details).

The committee established early in Chapter 1 a vision of some of the characteristics that might describe a "Resilient Nation in 2030." Using the information contained in this report, we expand upon this vision of characteristics of a "Resilient Nation in 2030" as part of the platform from which the vision and strategy for a resilient nation could be developed with leadership from the federal government (Box 8.1). The findings and five recommendations that follow Box 8.1 frame key actions that can help guide the nation in advancing collective, resilience-enhancing efforts to fulfill the national resilience vision the committee recommends be established.

BOX 8.1
Characteristics of a Resilient Nation in 2030

The nation, from individuals to the highest levels of government, has embraced a "culture of resilience." Information on risks to and vulnerability of individuals and communities is transparent and easily accessible to all. Proactive investments and policy decisions, including those for preparedness, mitigation, response, and recovery, have reduced the loss of lives, costs, and socioeconomic impacts of disasters. Community coalitions are widely organized, recognized, and supported to provide essential services before and after disasters occur. Recovery after disasters is rapid and includes funding from private capital. The per-capita federal cost of responding to disasters has been declining for a decade.

Key elements of this culture of resilience include

- Individuals and communities realize that they provide their own first line of defense against disasters.
- National leadership in resilience is implemented by policy decisions, funding, and actions throughout all federal agencies and Congress.
- Federal, state, and regional investment in and support for community-led resilience efforts are pervasive.
- Site-specific information on risk is readily available, transparent, and effectively communicated. This information has triggered dialogue within communities regarding the risks they face and how best to actively prepare for and manage them.
- Based on risk information, zoning ordinances are enacted and enforced that protect critical functions and help communities reap the benefit of natural defenses to natural hazards (e.g., floodplains, coastal wetlands, sand dunes).

- Building codes and retrofit standards have been widely adopted and are strictly enforced.
- A significant proportion of post-disaster recovery is funded through private capital and insurance payouts.
- Insurance premiums are risk based, and private insurers provide substantial premium reductions for buildings meeting current codes or retrofit standards.
- To speed recovery, community coalitions have developed contingency plans for governance and business continuity as well as for providing services, particularly for the most vulnerable populations.
- Post-disaster recovery is greatly accelerated by sufficient redundancy in infrastructure upgraded and hardened to take into account regional interdependencies.

Also included in these characteristics of a resilient nation (but well beyond the scope of recommendations) are a vibrant and diverse economy and citizenry who are safer, healthier, and better educated than previous generations.

The five recommendations below recognize that achieving resilience requires efforts and actions by individuals, families, communities, all levels of government, the private sector, academia, and community-based organizations including the nonprofit and faith-based groups. The process for improving resilience is dynamic, adaptive, and transparent and acknowledges the existence of interconnected and interdependent sets of social, economic, natural, and manmade systems that support communities. Recognition that events and their consequences do not adhere to geopolitical borders is also important. Embedded in each recommendation is also the need to continue long-term, prudent science and technology resilience research innovations.

The recommendations recognize that while physical resilience is a foundation, human resilience is the engine that drives the ability to absorb, recover from, and adapt to adverse events. No single sector or entity has ultimate responsibility for creating the foundation and driving the engine of resilience. These are shared responsibilities.

Risk Management and Reduction (from Chapters 2 and 5)

Finding: A variety of complementary structural and nonstructural measures exist to manage disaster risk. Risk management is, at its foundation, a community decision, and the risk management approach will only be effective if community members commit to using the risk management tools and measures available. Examples from actual disasters and their aftermaths, such as the June 2008 flood in Cedar Rapids, show that implementation of risk management strategies involves a combination of actors in local, state, and federal government, nongovernmental organizations (NGOs), researchers, the private sector, and individuals in the neighborhood community. Each will have different roles and responsibilities in developing the risk management strategy and in characterizing and implementing measures or tools, whether structural or nonstructural, to be added to the community's risk management portfolio. Some strategies can be implemented over the short term, whereas others may take a longer time.

Recommendation 2: **The public and private sectors in a community should work cooperatively to encourage commitment to and investment in a risk management strategy that includes complementary structural and nonstructural risk-reduction and risk-spreading measures or tools.**

The portfolio of tools should seek equitable balance among the needs and circumstances of individuals, businesses, and government, as well as the community's economic, social, and environmental resources. Among the most promising actions that would achieve results are in the areas of building codes and standards, and insurance.

Steps for Implementation:

Federal agencies, together with local and regional partners, researchers, professional groups, and the private sector can develop an essential framework (codes, standards, and guidelines) that drives the critical structural functions of resilience. Furthermore, cooperative work between the public and private sectors can encourage investment in nonstructural risk reduction measures such as insurance premiums; such premiums can include multiyear policies tied to the property with premiums reflecting risk. Specific focus on (a) building codes and standards and (b) insurance carry promise toward implementing this recommendation.

Finding 2a: Building codes and standards are effective in mitigating and reducing disaster risk to communities. For example, research and practice have demonstrated the value of building new homes to code and to increased standards in areas that may experience high winds or hurricanes. Of 13 homes built to a Fortified standard (Fortified standard is an increased building standard—above regular code—developed by the Institute for Business and Home Safety) on the Bolivar Peninsula, Texas, before Hurricane Ike, 10 survived that disaster. However, codes and standards have some variability due to the nature of local hazards; across the nation, codes and standards are unevenly enforced and many people do not know they exist. In addition to codes and standards, guidelines, certifications, and practices can also be effective in fostering resilience.

Recommendation 2a: **Federal agencies, together with local and regional partners, researchers, professional groups, and the private sector should develop an essential framework (codes, standards, and guidelines) that drive the critical structural functions of resilience.** This framework should include national standards for infrastructure resilience and guidelines for land use and other structural mitigation options, especially in known hazard areas such as floodplains. The Department of Homeland Security is an appropriate agency to help coordinate this government-wide activity. The adoption and enforcement of this framework at the local level should be strongly encouraged by the framework document.

Finding 2b: Investments in risk-spreading or risk-reducing measures through insurance and other financial instruments can facilitate mitigation, including the relocation of businesses, residences, and infrastructure out of hazard-prone areas. Vouchers given to lower-income property owners currently residing in hazard-prone areas could allow these property owners to afford all-hazards insurance; home improvement loans could be used to spread the upfront cost of risk reduction and mitigation measures over time; and seals of approval could be used to show that the property meets mitigation standards, thus enhancing its potential resale value.

Recommendation 2b: **The public and private sectors should encourage investment in risk-based pricing of insurance in which insurance premiums are designed to include multiyear policies tied to the property with premiums reflecting risk.** Such risk-based pricing reduces the need for public subsidies of disaster insurance. Risk-based pricing clearly communicates to those in hazard-prone areas the different levels of risk that they face. Use of risk-based pricing could also reward mitigation through premium reductions and could apply to both privately and publicly funded insurance programs.

National Disaster Loss Data (Chapter 3)

Finding: The ability to measure and evaluate the assets of communities and to understand the economic and human value of resilience is critical to improving disaster resilience. Because the assets of a community involve more than the high-value essential assets such as hospitals and utilities, but also include other resources with high social, cultural, and environmental value, decision-making models developed by communities have to involve both quantitative and qualitative "valuation" of assets in order to prioritize resilience investments.

In developing the case for enhancing resilience now and providing motivation for community decision makers to understand their inventory of assets and the ways in which they interact with one another, the historical spatial and temporal patterns of economic and human disaster losses on communities in the United States is important. Although the data available to assess economic and human losses nationally are conservative and are neither comprehensive nor centrally archived for the nation, the historical patterns of economic losses from hazards and disasters in the United States appear to be increasing and will be difficult to absorb, if allowed to continue. Without an all-hazards national repository for hazard event and loss data, estimates of how much or where losses are increasing or decreasing are difficult to make with any degree of statistical confidence. This lack of data compromises the ability of communities to make informed decisions about resilience-building strategies.

Recommendation 3: **A national resource of disaster-related data should be established that documents injuries, loss of life, property loss, and impacts on economic activity. Such a database will support efforts to develop more quantitative risk models and better understand structural and social vulnerability to disasters.** To improve access to these data, the principle of open access should be recognized in all relevant federal data management policies. The data should be made accessible through an Internet portal maintained either by a designated agency or by an independent entity such as a university. The National Science and Technology Council (NSTC) in the White House would be an appropriate entity to convene federal and state agencies, private actors, NGOs, and the research community to develop strategies and policies in support of these data-collection and maintenance goals.

Steps for Implementation:

(a) NSTC, or a federal body with a similar capacity, could convene federal agencies, private actors, and the research community to improve post-event data collection and public access to such data. Likely federal actors include the Federal Emergency Management Agency, National Oceanic and Atmospheric Administration, Centers for Disease Control and Prevention, U.S. Geological Survey, U.S. Forest Service, U.S. Department of Agriculture, and U.S. Army Corps of Engineers.

(b) Federal agencies, together with the private sector and research community, could determine essential data, standards, and protocols to employ, and which agencies are best positioned to collect and archive specific data on the impacts of hazards. Such an approach helps to avoid duplication of efforts.

(c) Biennial status reports coordinated by the NSTC on the nation's resilience could be based on analysis of these data and could include priorities for future data collection and dissemination.

National Resilience Scorecard (Chapter 4)

Finding: Without some numerical basis for assessing resilience, it would be difficult to monitor changes or show that community resilience has improved. At present, no consistent basis for such measurement exists.

***Recommendation 4:* The Department of Homeland Security in conjunction with other federal agencies, state and local partners, and professional groups should develop a National Resilience Scorecard.**

Steps for Implementation:

(a) General considerations:

- The scorecard should be readily adaptable to the needs of communities and levels of government, focusing specifically on the hazards that threaten each community.
- The scorecard should not attempt unreasonable precision, either in the ways in which individual factors are measured, or in the ways they are combined into composite indicators. Rather, qualitative and quantitative measures should be mingled, and reduced where appropriate to ordinal (rankings) rather than interval or ratio scales.

(b) Specific dimensions of the scorecard might include

- Indicators of the ability of critical infrastructure and businesses to recover rapidly from impacts;
- Social factors that enhance or limit a community's ability to recover, including social capital, language, and socioeconomic status;
- Indicators of the ability of buildings and other structures to withstand earthquakes, floods, severe storms, and other disasters;
- Indicators of the ability of businesses and markets to recover; and
- Factors that capture the special needs of individuals and groups, related to minority status, mobility, or health status.

Support and Establish Community Coalitions (Chapter 5)

Finding: Resilience requires reinforcement of our physical environment—the buildings and critical infrastructure that constitute the communities in which people live. It also requires the strengthening of the nation's social infrastructure—the local community networks that can mobilize to plan, make decisions, and communicate effectively. The principal action through which a local community could vastly accelerate progress toward enhanced resilience of its social and physical infrastructure is the establishment of a problem-solving coalition of local leaders from public and private sectors, with ties to and support from federal and state governments, and with input from the broader citizenry. The charge of such a coalition is to assess the community's exposure and vulnerability to risk, educating and communicating about risk, and evaluating and expanding its capacity to handle such risk. A truly robust coalition has at its core a strong leadership and governance structure, with a person or persons with adequate time, skill, and dedication necessary for the development and maintenance of relationships among all partners.

Recommendation 5: **Federal, state, and local governments should support the creation and maintenance of broad-based community resilience coalitions at local and regional levels.** Such coalitions can help communities promulgate and implement the proposed national resilience standards and guidelines for communities, and to assist them in the development and completion of the proposed National Resilience Scorecard.

Steps for Implementation:

(a) Assessment by the Department of Homeland Security and the Department of Health and Human Services—to the extent that these two agencies administer state and local grant programs to bolster national preparedness capabilities—of present federal funding frameworks and technical guidance. Such an assessment could gauge whether communities have sufficient support and incentive to adopt collaborative problem-solving approaches toward disaster resilience and emergency management.

(b) Adoption by communities of collaborative problem-solving approaches in which all private and public stakeholders (e.g., businesses, NGOs, community-based organizations, and faith-based organizations) are partners in identifying hazards, developing mitigation strategies, communicating risk, contributing to disaster response, and setting recovery priorities. The emergency management community is an integral part of these discussions, with potential to take a leadership role.

(c) Commitment by state and local governments to ensure that modern zoning laws and building codes are adopted and enforced.

(d) Commitment by state and local governments to secure adequate personnel to create and sustain public–private resilience partnerships, to promulgate and implement proposed national resilience standards and guidelines for communities, and to assist communities in the completion of the proposed national resilience scorecard.

Federal Policy Review (Chapter 6)

Finding: The development of appropriate policies, creation of optimal governance structures, and informed and coordinated management at all levels of government are crucial to improving community resilience. Community resilience will grow as the knowledge, experience, and understanding of these roles and responsibilities grow among decision makers at all levels of government.

Currently, a multitude of activities, programs, and policies exist at local, state, and federal levels to address some part of resilience for the nation. Several of the critical processes, such as land-use planning and building code enforcement, are the responsibility of local groups or governments. The federal policy role is primarily to ensure that resilience policies are nationally consistent and to provide information and best practices for development of appropriate policies at all levels. Consideration of potential unintended consequences of a new policy with respect to disaster resilience is also important.

The nation does not have an overall vision or coordinating strategy for resilience. Recent work on homeland security and disaster reduction are good beginnings, but the current suite of policies, practices, and decisions affecting resilience are conducted on an ad hoc basis with little formal communication, coordination, or collaboration. In fact, some policies, decisions and practices actually erode resilience.

Leaders at the local, state, and federal level are increasingly aware of community resilience and how it might be advanced through a variety of decisions and processes. Although many of those critical decisions and processes to improve resilience occur at the state and local levels, the federal government plays a central role in providing guidance for policy and program development to assist local communities in their pursuit of greater resilience. Development of new policies informed by an awareness of resilience, how it can be promoted through decisions and processes, and how resilience can be unintentionally eroded through poorly informed decisions is essential.

***Recommendation 6*: All federal agencies should ensure that they are promoting and coordinating national resilience in their programs and policies. A resilience policy review and self-assessment within agencies and strong communication among agencies are keys to achieving this kind of coordination.**

Steps for Implementation:

This commitment will require that each federal agency conduct a resilience self-assessment and communicate the analysis of its key resilience programs and activities to agency staff, to key partners and stakeholders, and to the public. Such an assessment includes

(a) The manner in which each agency's mission contributes to the resilience of the nation;
(b) How an agency's programs provide knowledge or guidance to state and local officials for advancing resilience;

(c) Evaluation by each federal agency of its interactions with other federal agencies, state and local governments, and the public to evaluate the extent to which its resilience work is made available to those who need it;
(d) Evaluation across federal agencies engaged in disaster services regarding what is working and what is not working, and
(e) Participation by each relevant federal agency in the coordination of resilience policy and programs as prescribed in PPD-8.

Appendix A

Committee and Staff Biographical Information

COMMITTEE BIOGRAPHIES

Susan L. Cutter, *Chair,* is a Carolina Distinguished Professor of Geography at the University of South Carolina, and director of the university's Hazards and Vulnerability Research Institute. Her primary research interests are in the area of vulnerability/resiliency science—what makes people and the places where they live vulnerable to extreme events and how vulnerability and resilience are measured, monitored, and assessed. She has authored or edited 12 books and more than 100 peer-reviewed articles and book chapters. Dr. Cutter has also led post-event field studies of the role of geographic information technologies in rescue and relief operations in the September 11th World Trade Center attack and studies of evacuation behavior from Three Mile Island (1979), Hurricane Floyd (1999), and the Graniteville, South Carolina, train derailment and chlorine spill (2005). She led a Hurricane Katrina post-event field team to coastal Mississippi (2006) and since then has been studying the community differences in long-term recovery of the Mississippi coast. She has provided expert testimony to Congress on hazards and vulnerability and was a member of the U.S. Army Corps of Engineers Interagency Performance Evaluation Taskforce that evaluated the social impacts of the New Orleans and Southeast Louisiana Hurricane Protection System in response to Hurricane Katrina. She has authored a Trends and Outlook report for the U.S. Army Corps of Engineers on Natural and Human-Induced Disasters and Other Factors Affecting Future Emergency Response and Hazard Management. Dr. Cutter serves on many national advisory boards and committees, including those of the National Research Council, American Association for the Advancement of Science, National Science Foundation, Natural Hazards Center, and the American Geophysical Union. She is a member of the International Council for Science's Integrated Research on Disaster Risk Scientific Committee. She was a coordinating lead author of Chapter 5 of the Intergovernmental Panel on Climate Change Special Report on "Managing the Risks of Extreme Events and Disasters to Advance Climate Change Adaptation." Dr. Cutter serves as co-executive editor of *Environment* and is an associate editor of *Weather, Climate, and Society*. She is a fellow of the American Association for the Advancement of Science and past president of the Association of American Geographers (2000) and past president of the Consortium of Social Science Associations (2008). In 2006, Dr. Cutter was the recipient of the Decade of Behavior Research Award given by a multidisciplinary consortium of more than 50 national and international scientific organizations in the social and behavioral sciences. In 2011, she received the Lifetime Achievement Award from the Association of American Geographers. Dr. Cutter holds the Munich RE Foundation Chair (2009-2012) on Social Vulnerability through the United Nations University–Institute for Environment and Human Security, in Bonn, Germany. She received her B.A. from California State University, East Bay and her M.A. and Ph.D. from the University of Chicago.

Joseph A. "Bud" Ahearn (member, National Academy of Engineering) is a recently retired senior executive at CH2M HILL, where he was an executive leader in the engineering business lines of transportation, environment, water, industrial design, and related infrastructure. During his 18-year career at CH2M HILL, he served as vice chairman of the board with responsibilities for strategic planning, governmental affairs, strategic communications, and leadership development, and also served in several other capacities including Transportation Business Group president, Eastern Region manager, senior vice president, Federal Programs director, and principal-in-charge for two major transportation corridor projects in California. Prior to joining CH2M HILL, Mr. Ahearn had a distinguished military career spanning three decades, where he achieved the rank of Major General in the U.S. Air Force. During his 34 years with the Department of Defense, General Ahearn was responsible for shaping financial strategy, developing budgets, and executing infrastructure programs totaling more than $7 billion annually. As the senior civil engineer for the U.S. Air Force, he directed the operational readiness and natural disaster response of U.S. Air Force combat engineers and the development and operations of all U.S. air bases around the world. Dedicated to advancing engineering education and providing sustainable systems and services in developing countries, he is a founding sponsor and governing board director of Engineers Without Borders USA. Committed to strengthening the engineering profession, he is an active Distinguished Member of the American Society of Civil Engineers (ASCE), ASCE's Industry Leaders Council, the National Academy of Engineering, and the National Academy of Construction. In addition to numerous military awards, General Ahearn received the Air Force Order of the Sword, the highest honor that the Noncommissioned Officer Corps of the U.S. Air Force can bestow; the University of Notre Dame College of Engineering Honor Award for professional achievement; and the Newman Medal from the Society of American Military Engineers (SAME) for outstanding military engineering achievement in Europe. He was also recipient of the SAME Golden Eagle award for lifetime achievement and was named an honorary member of the American Institute of Architects.

Bernard Amadei (member, National Academy of Engineering) is professor of civil engineering at the University of Colorado, Boulder. Dr. Amadei's research interests cover the topics of sustainability, system dynamics, and international development. At the university, he directs the Mortenson Center in Engineering for Developing Communities that has an overall mission to educate globally responsible engineering students and professionals to offer sustainable and appropriate solutions to the endemic problems faced by developing communities. His research at the University of Colorado has been multidisciplinary. He has also provided consulting services to various engineering companies and organizations around the world. Dr. Amadei is also the founding president of Engineers Without Borders USA and the cofounder of Engineers Without Borders International network. He has coauthored several books and approximately 160 technical papers. Among other distinctions, Dr. Amadei is the 2007 co-recipient of the Heinz Award for the Environment; the recipient of the 2008 ENR Award of Excellence; an elected member of the U.S. National Academy of Engineering; and an elected Senior Knight-Ashoka Fellow. He received his M.S. degree in civil engineering from the University of Toronto and his Ph.D. in civil engineering from the University of California, Berkeley. In addition, he holds three honorary doctoral degrees.

Patrick Crawford coordinates disaster preparedness and relief efforts for the Feeding America Network. His responsibilities include collaborating with national partners in the emergency management and nonprofit communities to ensure effective collection and distribution of donated food items following disaster. Mr. Crawford also directs internal operations during disaster by coordinating among several Feeding America departments including Logistics, Food Sourcing, Philanthropy, Communications, and Government Relations and directly with the over 200 food banks throughout the network. Prior to joining Feeding America, Mr. Crawford served as the Director of the Midwest Region for James Lee Witt Associates (JLWA), a crisis and consequence management firm, where he led efforts in emergency preparedness, response, recovery, and mitigation and provided strategic counsel and government relations advice to mitigate future flood losses near the Mississippi and Missouri Rivers. Mr. Crawford worked for over 16 years with the Federal Emergency Management Agency (FEMA) in crisis and consequence management, including responses to floods, earthquakes, hurricanes, wildfires, and acts of terrorism. At FEMA he worked extensively in the Gulf region, following the catastrophic Hurricanes Katrina and Rita, and in FEMA's Region 9 (covering California, Arizona, Nevada, Hawaii, and U.S. territories in the Pacific), where he worked directly with state and local governments to build emergency response, recovery, mitigation, and preparedness capacity. Mr. Crawford served as a Captain in the U.S. Army Chemical Corps where his primary responsibility was to ensure unit readiness with regard to Nuclear, Biological, and Chemical defense activities. He received his B.A. in government from the University of Notre Dame and his M.A. in education from Loyola College of Maryland.

Gerald E. Galloway, Jr. (member, National Academy of Engineering) is the Glenn L. Martin Institute Professor of Engineering and an affiliate professor of public policy at the University of Maryland, College Park. His 38-year career in the military included positions such as commander of the Army Corps of Engineers District in Vicksburg, Mississippi, and professor and founding head of the Department of Geography and Environmental Engineering and dean of the Academic Board at the U.S. Military Academy. He was promoted to brigadier general in 1990 and retired from active duty in 1995. A civil engineer, public administrator, and geographer, Dr. Galloway's current research focuses on the development of U.S. national water policy in general and national floodplain management policy in particular. He is currently a member of the National Research Council's Water Science and Technology Board and the Disasters Roundtable. A member of the National Academy of Engineering, Dr. Galloway earned his M.S.E. at Princeton and his Ph.D. in geography (specializing in water resources) from the University of North Carolina at Chapel Hill.

Michael F. Goodchild (member, National Academy of Sciences) is a professor of geography and director of the Center for Spatial Studies and Center for Spatially Integrated Social Science at the University of California, Santa Barbara. He is also chair of the Executive Committee of the National Center for Geographic Information and Analysis and associate director of the Alexandria Digital Library. He taught at the University of Western Ontario for 19 years before moving to his present position in 1988. His research interests focus on the issues of geographic information, including accuracy and the modeling of uncertainty, the design of spatial decision support systems, the development of methods of spatial analysis, and data structures for global geographic information systems. He has explored using digital information gathered by remote sensing satellites to create spatial and environmental models of the planet, make maps, and create digital libraries of geographic information that can be widely accessed electronically. He has also developed mathematical models to help quantify the difference between these geographic measurements and the reality of the world outside, so that geographic information can be accurately used. His research also includes digital libraries and problems associated with search, retrieval, and use of geographic information over the Internet; the potential for novel kinds of fieldwork enabled by fully mobile, wirelessly connected, and even wearable information technology; and the role of geographic information technologies in science and policy making. He has received several awards and published numerous books and journal articles. A member of the National Academy of Sciences, he has served on numerous National Research Council study and standing committees as both member and chair. He received a B.A. in physics from Cambridge University and a Ph.D. in geography from McMaster University.

Howard C. Kunreuther is the James G. Dinan Professor of Decision Sciences and Public Policy and codirector of the Risk Management and Decision Processes Center at the Wharton School, University of Pennsylvania. He has a longstanding interest in ways that society can better manage low-probability, high-consequence events related to technological and natural hazards. Dr. Kunreuther is a fellow of the American Association for the Advancement of Science (AAAS) and recently served as a member of the National Academy of Sciences Panel on Adaptation Strategies for Climate Change. He is a Distinguished Fellow of the Society for Risk Analysis, receiving the Society's Distinguished Achievement Award in 2001. Dr. Kunreuther is a member of the World Economic Forum's Global Agenda Council on Insurance and Asset Management for 2011-2012, and in 2009-2010 served as cochair of the Forum's Global Agenda Council on Leadership and Innovation for Reducing Risks from Natural Disasters. He currently serves the Intergovernmental Panel on Climate Change (IPCC) as a chapter lead author of the IPCC's 5th Assessment Report, *Integrated Risk and Uncertainty Assessment of Climate Change Response.* His most recent books are *Learning from Catastrophes: Strategies for Reaction and Response* (with M. Useem) (2010), and *At War with the Weather* (with E. Michel-Kerjan) (2009; paperback, 2011), winner of the Kulp-Wright Book Award from the American Risk and Insurance Association in 2011. He received his A.B. in economics from Bates College and his Ph.D. in economics from Massachusetts Institute of Technology.

Meredith Li-Vollmer is a risk communication specialist for Public Health–Seattle & King County, where she leads planning for communications during emergencies, with a particular focus on strengthening the capacity of public health to reach those most at risk during emergencies. In this role, she conducts audience research, directs public engagement projects, develops strategies and materials for public outreach, and serves in a public information role during emergency activations. Meredith is also a researcher with the University of Washington Preparedness & Emergency Response Research Center and a clinical assistant professor at the University of Washington School of Public Health and Community Medicine. Recent bodies of work include public engagement projects on crisis standards of care and vaccine distribution, research on text messaging for public health emergencies, and development of comic books about disaster survivors. Her work has received multiple awards, including the Model Practice Award from the National Association of City and County Health Officials and the Gold Award for Excellence from the National Public Health Information Coalition. Prior to joining Public Health–Seattle & King County, Meredith taught communications at the University of Washington. She received her Ph.D. in communications from the University of Washington.

Monica Schoch-Spana, a medical anthropologist, is a senior associate with the Center for Biosecurity of the University of Pittsburgh Medical Center (UPMC) and an assistant professor in the School of Medicine Division of Infectious Diseases. The Center for Biosecurity works at the intersection of public health and national security, to affect policy and practice in ways that improve U.S. resilience to biological and nuclear dangers. Since 1998, Dr. Schoch-Spana has briefed numerous federal, state, and local officials, as well as medical, public health, and public safety professionals on critical issues in biosecurity and public health emergency preparedness. National advisory roles include serving on the Steering Committee of the Disasters Roundtable of the National Research Council (NRC), the Institute of Medicine Standing Committee on Health Threat Resilience, and the NRC Committee to Review the Department of Homeland Security's Approach to Risk Analysis. In particular, she has led research, education, and advocacy efforts to encourage greater consideration by authorities of the public's key contributions to the management of epidemics, biological attacks, and other public health emergencies. In 2009, she organized the national conference Resilient American Communities: Progress in Policy and Practice, and chaired the Resilience Research Work Group. In 2006, she oversaw the Working Group on Citizen Engagement in Health Emergency Planning, and was the principal organizer for the U.S.-Canada summit on Disease, Disaster & Democracy—The Public's Stake in Health Emergency Planning. In 2003, she organized the national meeting, Leadership During Bioterrorism: The Public as an Asset, Not a Problem, and chaired the Working Group on "Governance Dilemmas" in Bioterrorism Response that issued consensus recommendations to mayors, governors, and top health officials nationwide in 2004. She serves on the faculty for the National Consortium for the Study of Terrorism and Responses to Terrorism (START), a university-based center of excellence supported by the U.S. Department of Homeland Security. In 2003, Dr. Schoch-Spana helped establish the Biosecurity Center of UPMC; prior to that she worked at the Johns Hopkins Center for Civilian Biodefense Strategies starting in 1998. She received her Ph.D. in Cultural Anthropology from Johns Hopkins University and B.A. from Bryn Mawr College.

Susan C. Scrimshaw (member, Institute of Medicine) is president of The Sage Colleges. She moved to Sage after serving as president of Simmons College. Dr. Scrimshaw was formerly dean of the University of Illinois at Chicago (UIC) School of Public Health and professor of community health sciences and anthropology at UIC. Under her leadership, the UIC School of Public Health established a wide range of community, regional, and national partnership initiatives, including addressing disparities in the delivery of health care, improving pregnancy outcomes, maternal and child health, healthy aging, violence prevention, cancer prevention, AIDS/STD prevention, and occupational and environmental health issues. While dean of the School of Public Health, she led the school in a national role in responding to the September 11 terrorist attacks. Her own interdisciplinary research has focused on gender, race, ethnicity, and culture, and their impact on public health and includes community participatory research methods, addressing health disparities, improving pregnancy outcomes, violence prevention, health literacy, and culturally appropriate delivery of health care. She has been frequently honored for her work in raising awareness of public health issues around the world, including minority populations in the United States. Her awards include a gold medal as a "Hero of Public Health," presented by the president of Mexico, and the Margaret Mead Award of the American Anthropological Association and the Society for Applied Anthropology. She is the author of five books or monographs and numerous journal articles and book chapters. She is a fellow of the American Association for the Advancement of Science, past president of the Society for Medical Anthropology, past chair of the national Association of Schools of Public Health, and served on the board of directors and as chair of the U.S.-Mexico Foundation for Science. She was a founding member of the task force on Community Preventive Services of the federal Centers for Disease Control and Prevention. In 2006, she was awarded the Illinois Public Health Association's highest honor, the 2006 Distinguished Service Award, in recognition of her distinguished service in research, teaching, and public health practice. She served on the governing council of the Institute of Medicine (IOM) of the National Academies and on the National Research Council (NRC) Committee on Science, Engineering, and Public Policy, as well as many IOM and NRC panels and boards. The Albany/Colonie Chamber of Commerce recently named her a Woman of Influence in the Distinguished Career category. Dr. Scrimshaw received an A.B. from Barnard College and a Ph.D. in anthropology from Columbia University.

Ellis Stanley is a vice president for Emergency Management Services at Dewberry LLC. Prior to joining Dewberry, Ellis served as general manager of the City of Los Angeles Emergency Preparedness Department. Before that, he was director of the Atlanta-Fulton County Emergency Management Agency. In 2008 he served as director of Democratic National Convention planning for the City and County of Denver, Colorado. With more than 35 years of experience in the emergency management field, Ellis has worked at four national political conventions, the 1996 Olympic Games in Atlanta, and the 1994 Papal visit and World Youth Conference in Denver. He is currently serving on the Board of Directors of the Greater Los Angeles Red Cross Chapter and chairs the Response Committee. He served as chair of the Emergency Management Accreditation Program and the Board of Directors of Operation Hope and the Disaster Recovery Institute International. Ellis is a past president of the International Association of Emergency Managers and has led delegations of emergency management professionals to China, Japan, and other countries. He is currently a member of the International Association of Emergency Managers Global Board of Directors. Ellis serves as an adjunct professor at American University teaching Senior Crisis Management and at Harvard University teaching Meta-Leadership. He is currently chair of the National Research Council's Disasters Roundtable. He was elected a Fellow of the National Academy of Public Administration in 2007 and inducted into Contingency Planning and Management Hall of Fame's Public Servant in

2005. Ellis graduated from the University of North Carolina at Chapel Hill in 1973 with a degree in political science. He is a graduate of the Executive Leadership Program for Senior Homeland Security Officials for the Post Naval Graduate School in Monterey, California, and a graduate of the John F. Kennedy School of Government's National Preparedness Leadership Initiative. Ellis was awarded an Honorary Doctor of Public Service degree by the University of Maryland Eastern Shore in 2009.

Gene Whitney recently retired as Energy Research Manager for the Congressional Research Service at the Library of Congress in Washington, D.C. Previously, he was assistant director for Environment at the White House Office of Science and Technology Policy (OSTP). His work at OSTP focused on the science and technology policy aspects of earth sciences, natural hazards and disasters, energy, water, land remote sensing, environment, and natural resources. He served as cochair of the U.S. Group on Earth Observations and was OSTP liaison to the U.S. Climate Change Science Program. He directed the Future of Land Imaging Interagency Working Group, and served as National Science Technology Council director for the Subcommittee on Disaster Reduction and the Subcommittee on Water Availability and Quality. Dr. Whitney coordinated the federal interagency science and technology portfolio for the United States in UNESCO. He served as a member of the Joint U.S.–Canada Task Force investigating the massive electrical blackout of August 14, 2003, in the northeastern United States and southern Canada, and worked with the President's Council of Advisors on Science and Technology on national energy efficiency policy. Prior to OSTP, Dr. Whitney was chief scientist for the U.S. Geological Survey Energy Resources Team, where he managed the energy research and assessment group, conducting basic research on the geology, geochemistry, and geophysics of fossil fuels; conducting national and global assessments of oil, natural gas, and coal resources; and assessing availability and economics of fossil fuels. He has authored or coauthored numerous scientific papers and abstracts. He received a National Research Council postdoctoral fellowship at the National Aeronautics and Space Administration/Jet Propulsion Laboratory and was awarded a senior postdoctoral fellowship at Ecole Normale Superieur in Paris. His international experience includes working with the governments of China, Russia, Pakistan, Algeria, Bangladesh, and Japan on energy and mineral resource issues. Dr. Whitney received his Ph.D. in geology from the University of Illinois.

Mary Lou Zoback (member, National Academy of Sciences) is a seismologist and consulting professor in the Environmental Earth System Science Department at Stanford University. From 2006 to 2011 she was vice president for Earthquake Risk Applications with Risk Management Solutions, a private catastrophe modeling firm serving the insurance industry. She was previously a senior research scientist at the U.S. Geological Survey (USGS) in Menlo Park, California, where she served as chief scientist of the Western Earthquake Hazards team. Dr. Zoback has served on numerous national committees and panels on topics including defining the next generation of Earth observations from space, storage of high-level radioactive waste, facilitating interdisciplinary research, and science education. She is a member of the U. S. National Academy of Sciences, past president of the Geological Society of America, and past chair of both the Southern California Earthquake Center Advisory Council and the Advisory Committee for San Francisco's Community Action Plan for Seismic Safety program. She is currently a member of the National Research Council's Disasters Roundtable. She joined the USGS in 1978 after receiving her B.S., M.S., and Ph.D. in geophysics from Stanford University.

STAFF BIOGRAPHIES

Lauren Alexander Augustine is the Associate Executive Director in the Division on Earth and Life Studies and Director of the Disasters Roundtable at The National Academies. Dr. Alexander Augustine also serves as the Country Director in the Academies' African Science Academy Development Initiative. She came to the National Academies in 2002 as a study director for the Water Science and Technology Board in the National Research Council and directed many studies on a range of water resources topics, including Texas instream flows, endangered species in the Klamath and Platte River Basins, and forest hydrology. Previously, Dr. Alexander Augustine worked at the U.S. Geological Survey, Water Resources Division, doing hydrogeomorphic research in Coastal Plain wetlands. Dr. Alexander Augustine received her B. S. in applied mathematics and systems engineering and her Masters degree in environmental planning and policy from the University of Virginia; she completed her Ph.D. from Harvard University in an interdisciplinary program that combined physical hydrology, geomorphology, and ecology.

John H. Brown, Jr. is the Program Associate for the Disasters Roundtable at the National Academies in the Division of Earth and Life Studies. He came to the Academies in 2002 and has worked on numerous studies in conjunction with the Board on Environmental Studies and Toxicology, including toxicity pathway-based risk assessment, the hidden costs of energy, a research and restoration plan for Western Alaska salmon, risk reduction and economic benefits from controlling ozone air pollution, and the environmental impacts of wind energy projects. Prior to joining the Academies staff, he worked with the Smithsonian Institution and the Kennedy Center. He received his bachelor's and master's degrees from Boston University.

Eric J. Edkin is a Senior Program Sssistant with the Board on Earth Sciences and Resources. He's background is in website and graphic design and began working for the National Academies in 2009. He supports the work of several standing committees of the Board and has worked on numerous studies including spatial data infrastructure, geospatial intelligence, earthquake engineering and earthquake resilience, and deep time and climate.

Elizabeth A. Eide is Director of the Board on Earth Sciences and Resources at the NRC and served as study director for this study. Prior to joining the NRC as a senior program officer in 2005, she was a researcher, team leader, and laboratory manager for 12 years at the Geological Survey of Norway in Trondheim. While in Norway her research included basic and applied projects related to isotope geochronology, mineralogy and petrology, and crustal processes around the world. Her publications include more than 40 journal articles and book chapters, and 10 Geological Survey reports. She has overseen 10 NRC studies and provided collaborative support for 4 others on topics covering energy and mineral resources, energy and mining workforce, induced seismicity, floodplain mapping, international geosciences, data and tools to address at-risk populations, Earth surface processes, and applied remote sensing applications. She completed a Ph.D. in geology at Stanford University and received a B.A. in geology from Franklin and Marshall College.

Neeraj P. Gorkhaly is the Research Associate for the Board on Global Science and Technology, and Committee on Science, Engineering, and Public Policy at the U.S. National Academies' Division of Policy and Global Affairs. Neeraj grew up on Kathmandu, Nepal and is a graduate of The Ohio State University. For the last seven years Neeraj has been in science policy area contributing to reports as *Rising Above the Gathering Storm, Responsible Science, Science and Technology for America's Progress, Facilitating Interdisciplinary Research* to name a few. Neeraj is an alumnus and was a mentor for the John Glenn Policy Internship Program in Washington DC. He is actively involved in the Nepalese-American community as an Ambassador for Grassroot Movement in Nepal (GMIN) and the founder of the Gorkhaly Foundation which organizes and administrates various programs and events for social awareness and development projects in Nepal.

Appendix B

Committee Meetings and Public Agendas

PUBLIC SESSIONS

FIRST COMMITTEE MEETING
Wednesday September 29, 2010
The Venable Conference Center
575 7th Street, NW, Washington, DC 20004-1601

Wednesday, September 29th

10:00-10:20 **Welcome and introductions**
Susan Cutter, chair

10:20-15:30 **Discussion with study sponsors**
Susan Cutter

Each sponsor has been asked to respond to four questions:
1. What is your agency's working definition of resilience?
2. What are your agency's expectations from this study?
3. Who are the target audiences for this study to meet those expectations?
4. What products should be produced from this study and for which audiences?

10:20-12:00 **Department of Homeland Security, Human Factors/Behavioral Sciences Division**
Michael Dunaway

DHS/Federal Emergency Management Agency
Keith Turi

U.S. Geological Survey
David Applegate and *Paula Gori*

	National Oceanic and Atmospheric Administration *Margaret Davidson* and *Keelin Kuipers*
	Community and Regional Resilience Institute *Warren Edwards* and *Heather Lair*
13:00-14:15	**Headquarters, U.S. Army Corps of Engineers, Directorate of Contingency Operations, International Emergency Management Program** *Andrew Bruzewicz*
	Office of Electricity Delivery and Energy Reliability, U.S. Department of Energy *Anthony Lucas*
	National Aeronautics and Space Administration *Craig Dobson*
	Department of Agriculture Forest Service *Mike Hilbruner*
14:30-15:30	**General discussion** *Susan Cutter*

WORKSHOP & SECOND COMMITTEE MEETING
Wednesday-Thursday, September 29-30, 2010
Astor Crowne Plaza Hotel
739 Canal Street at Bourbon, New Orleans

Tuesday, January 18th

Grand A Room

17:30-18:30 **Forgetting the Unforgettable: Social Memory and Resilience in New Orleans**
Craig Colten, Louisiana State University
Keynote presentation

Wednesday, January 19th

08:00-11:30 **Guided Tour of New Orleans**
Pam Jenkins, University of New Orleans
Doug Meffert, Tulane University

11:30-12:30 **Drive to Waveland, Mississippi**

12:30-15:00 **Visit stops in Waveland, Gulfport, and Biloxi, Mississippi**

15:00-17:00 **Discussion with Local Staff**

Discussion about resilience topics with
Alice Graham, Executive Director, Mississippi Coast Interfaith Disaster Task Force
John Hosey, Disaster Behavioral Health Project Manager, Mississippi Coast Interfaith Disaster Task Force

Discussion on Disaster Task Force
John Kelly, Chief Administrative Officer for the City of Gulfport
Rupert Lacy, Director, Harrison County Emergency Management Agency
Tom Lansford, Academic Dean and Professor, Political Science, University of Southern Mississippi

Discussion on Mississippi, Gulf Coast
Reilly Morse, Senior Attorney, Mississippi Center for Justice
Kimberly Nastasi, CEO, Mississippi Gulf Coast Chamber of Commerce
Tracie Sempier, Coastal Storms Outreach Coordinator, Mississippi-Alabama Sea Grant Consortium

Lori West, Gulf Region Director, IRD, U.S. Gulf Coast Community Resource Centers

Thursday, January 20th

Toulouse A & B Rooms

08:30-08:40 Welcome and introductions
Susan Cutter, Committee Chair

08:40-08:45 Opening Remarks
Senator Mary Landrieu, opening remarks (via videotape)

08:45-09:30 Keynote presentation: The New Orleans at Five: From Recovery to Transformation
Allison Plyer, Greater New Orleans Community Data Center

09:40-15:00 Panel Sessions
Facilitated by *Ann Olsen*, Meridian Institute

09:40-10:25 Business-Insurance-Real Estate Panel
Julie Rochman, President and CEO, Institute for Business & Home Safety
Eric Nelson, Travelers Vice President, Personal Insurance
Ommeed Sathe, Director of Real Estate Strategy, New Orleans Redevelopment Authority

10:40-11:25 Critical Infrastructure Panel
Marcia St. Martin, Executive Director, Sewerage and Water Board of New Orleans
Justin Augustine, CEO, New Orleans Regional Transit Authority and Vice President, Veolia Transportation
Greg Grillo, Entergy Corporation, Director–Transmission Project Management Construction and Incident Commander
Frank Wise, Executive Director of Network Operations, Florida, Verizon Wireless

11:30-12:15 Governance Panel
Earthea Nance, University of New Orleans
Bill Stallworth, Executive Director/Councilman East Biloxi Coordination and Relief Center/Biloxi City Council

Stephen Murphy, Director of Planning, City of New Orleans Office of Homeland Security and Emergency Response
Charles Allen, III, Advisor to the Mayor and Director, New Orleans Office of Coastal and Environmental Affairs

13:15-14:00 **Social Capital Panel**
Natalie Jayroe, CEO Greater New Orleans and Acadiana Food Bank
Steven Bingler, President, Concordia
Mary Claire Landry, Director, Domestic Violence Programs (Family Justice Center; Crescent House; Sexual Assault Services; and Project SAVE) Catholic Charities Archdiocese of New Orleans
Pam Jenkins, University of New Orleans

14:05-14:50 **Healthy Populations and Responsive Institutions Panel**
Joseph Donchess, Executive Director, Louisiana Nursing Home Association
Knox Andress, Designated Regional Coordinator, Louisiana Region 7 Hospital Preparedness; Louisiana State University Health Sciences Center-Shreveport, Louisiana Poison Center
Anthony Speier, Deputy Assistant Secretary, Office of Behavioral Health, Louisiana
Paul Byers, Acting State Epidemiologist, Mississippi State Department of Health

15:10-16:30 **Breakout Sessions**
Committee, Panelists, and Audience Discussion

WORKSHOP & THIRD COMMITTEE MEETING

Monday-Wednesday, March 7-9, 2010
The Hotel at Kirkwood Center
7725 Kirkwood Blvd. SW, Cedar Rapids, Iowa

Monday, March 7th

07:30-12:00 **Guided Tour of Cedar Rapids, IA**
Christine Butterfield, Director, Community Development, City of Cedar Rapids

Adam Lindenlaub, Long-Range Planning Coordinator, Corridor MPO, Cedar Rapids' Community Development Department
Dave Elgin, Director of Public Works, City of Cedar Rapids
Anne Strellner, St. Luke's Hospital
Julie Stephens, Linn County Public Health
Mike Goldberg, Linn County EMA
Drew Skogman, Skogman Realty
Steve Dummermuth, Downtown landowner

07:30-08:50 **West Cedar Rapids**

07:30-08:45 **Block by Block Headquarters and Vicinity with local staff**

08:50-12:00 **East Cedar Rapids**

09:05-09:45 **African-American Museum**

10:00-11:00 **Mercy Hospital**

11:00-12:00 **Small business/downtown commerce and properties**

12:00-12:40 **Iowa City and University of Iowa, Iowa Flood Center**

12:40-13:00 **C. Maxwell Stanley Hydraulics Laboratory**
Larry Weber, Director of IIHR-Hydroscience & Engineering and Co-organizer of the Iowa Flood Center

13:15-17:00 **Iowa Memorial Union Building, University of Iowa**
Discussion with faculty and students at campus

13:15-13:30 **Opening remarks on the immediate effects of the flood on campus**
Larry Weber, Director of IIHR-Hydroscience & Engineering and Co-organizer of the Iowa Flood Center

13:30-13:35 **Introductions to study and committee**
Susan Cutter, Committee chair

13:35-15:15 **University faculty/research pane discussions**
Moderated by *Gerry Galloway*, Committee member
Jerry Anthony, School of Urban and Regional Planning

Witold Krajewski, Department of Civil & Environmental Engineering and Director of the Iowa Flood Center
Kevin Leicht, Department of Sociology
Alan Macvey, Theatre Arts Department
Marizen Ramirez, Department of Occupational and Environmental Health
John Beldon Scott, School of Art and Art History
Kathleen Stewart, Department of Geography
Peter Thorne, Department of Occupational and Environmental Health
James Throgmorton, School of Urban and Regional Planning
Larry Weber, Department of Civil & Environmental Engineering and Director of IIHR—Hydroscience & Engineering
Michael Wichman, State Hygienic Laboratory

15:15-15:30 **Break**

15:30-17:00 **Student panel discussions: Facilitated discussion with 8-12 students, undergraduate and graduate, from University of Iowa (UI) and University of Northern Iowa (UNI)**
Moderated by *Meredith Li-Vollmer*, Committee member
Luciana Cunha, Civil & Environmental Engineering (UI)
Emily White, Geography (UI)
Achilleas Tsakiris, Civil & Environmental Engineering (UI)
Kimberly Hoppe, Occupational and Environmental Health (UI)
Maria Elisa Mandarim de Lacerda, Theater Arts (UI)
Amy Costliow, School of Health, Physical Education and Leisure Services (UNI)
Kari Dirksen, School of Health, Physical Education and Leisure Services (UNI)

Tuesday, March 8th

Ballroom B & C, Hotel at Kirkwood Center

08:30-08:40 Introductions
I, Committee chair

08:40-09:10 Opening remarks
Mayor Ron Corbett, Cedar Rapids
Jeff Pomeranz, City Manager, Cedar Rapids

09:15-15:30 Panel sessions

09:15-10:25 Private sector

Moderated by *Howard Kunreuther*, Committee member
Terri Vaughan, CEO, National Association of Insurance Commissioners
Jeff McClaran, Vice President, Incident Management, Central and Gulf States, Wells Fargo
Dee Brown, Director, Energy Delivery Operations Support, Alliant Energy

10:40-11:50 Government

Moderated by *Gene Whitney*, Committee member
Patty Judge, PJJ Solutions, Inc. (former Lt. Governor of Iowa)
Kamyar Enshayan, Cedar Falls City Council; Professor, University of Northern Iowa
Chuck Wieneke, City of Cedar Rapids City Council
Christine Butterfield, Community Development Director, City of Cedar Rapids

12:50-14:00 First and second responders

Moderated by *Patrick Crawford*, Committee member
Lt. General Ron Dardis, former Rebuild Iowa Office Executive Director and former Adjutant General of the Iowa National Guard
Clark Christensen, Logistics Officer, Public Health Response Team Coordinator, & State Coordinator for the Medical Reserve Corps, Iowa Department of Public Health, Center for Disaster Operations and Response
David Miller, Private consultant; former Administrator, Iowa Homeland Security and Emergency Management Division
Rick Wulfekuhle, Emergency Manager, Buchanan County
Mark English, Assistant Fire Chief, Cedar Rapids Fire Department

14:00-15:10 Community representatives

Moderated by *Susan Scrimshaw*, Committee member
Donna Harvey, Director, Iowa Department on Aging
Nancy Beers, Director of Disaster Services and Camp Noah Lutheran Social Service

Mitch Finn, Deputy Executive Director, Hawkeye Community Action Program
Bill Gardam, President and CEO, Horizons, A Family Service Alliance
Cindy Kaestner, Vice President/Executive Director, Abbe Center for Community Mental Health

---

WORKSHOP & FOURTH COMMITTEE MEETING
Tuesday, May 24, 2011
Beckman Center of the National Academies
100 Academy Way
Irvine, California

08:15-08:30 **Welcome and introductions**
Susan Cutter, Chair

08:30-09:30 **Keynote**
Laurie Johnson, Principal, Laurie Johnson Consulting | Research

09:30-10:45 **Infrastructure and economic recovery panel**
Moderated by *Mary Lou Zoback*, Committee member
John Holmes, Deputy Executive Director of Port Operations for the Port of Los Angeles
Chris Poland, Chairman, CEO, and Senior Principal, Degenkolb Engineers
Ezra Rapport, Executive Director of the Association of Bay Area Governments
Michael Morel, Manager of Operations and Planning for the Metropolitan Water District

11:00-12:15 **Risk communication and resilience indicators panel**
Moderated by *Monica Schoch-Spana*, Committee member
Roxane Silver, School of Social Ecology, University of California, Irvine
Sarah Karlinsky, Deputy Director of San Francisco Planning + Urban Research Association (SPUR)
Baruch Fischhoff, Departments of Social and Decision Sciences and of Engineering and Public Policy, Carnegie Mellon University

13:00-14:15 **Disaster communication and impacts panel**
Moderated by *Meredith Li-Vollmer*, Committee member
Lucy Jones, Chief Scientist for the Multi Hazards Initiative in Southern California, U.S. Geological Survey
Mariana Amatullo, Vice President, Designmatters at Art Center College of Design
Barbara Andersen, Strategic Partnerships Director, Orfalea Foundations

David Eisenman, UCLA Division of General Internal Medicine and Health Services Research

14:15-15:30 **IT/social media and disaster resilience panel**
Moderated by *Michael Goodchild*, Committee member
Leysia Palen, Department of Computer Science, University of Colorado, Boulder
Matt Zook, Department of Geography, University of Kentucky
Alan Glennon, Department of Geography, University of California, Santa Barbara
Nalini Venkatasubramanian, Department of Computer Science, University of California, Irvine

15:40-16:30 **Open plenary session**

WORKSHOP & FIFTH COMMITTEE MEETING
Wednesday, September 21, 2011
Keck Center of the National Academies
500 5th Street, NW
Washington, DC 20024

Wednesday, September 21st

Room 101
10:00-10:10 **Welcome and introductions**
Susan Cutter, Chair

10:10-13:20 **Presentations**
Paul Brenner, Senior Vice President, ICF International
Claire Rubin, Claire Rubin & Associates
Ben Billings, Senior Policy Adviser for Homeland Security and Disaster Recovery, Office of Senator Mary Landrieu

13:20-13:45 **Open discussion**

Appendix C

Essential Hazard Monitoring Networks

EARTHQUAKE AND VOLCANO MONITORING

The U.S. Geological Survey (USGS) Advanced National Seismic System (ANSS),[1] comprises federal, state, university, utility, and industry seismographic networks, provides near real-time (within minutes) information on the magnitude, location, and local shaking distribution for significant U.S. earthquakes. The USGS National Earthquake Information Center (NEIC) provides authoritative information on both U.S. and global earthquakes and is staffed 24 hours a day. The ANSS was authorized by Congress in 2002 to significantly upgrade and expand the nation's seismic monitoring capability; however, only 25 percent of the planned deployments had been achieved by the end of 2011 because of resource constraints. A recent National Research Council review of the multiagency National Earthquake Hazard Reduction Program noted that many of the review's other recommendations are critically dependent on data generated by ANSS (NRC, 2011a).

The USGS Volcano Hazard Program operates a monitoring network that includes local sensors (seismic, ground deformation, webcams, tilt, gas) combined with remote sensing on active volcanoes that pose a threat to lives, property, and air traffic (the latter through upper atmospheric ash clouds). Plans are currently under way to expand, modernize, and make interoperable the data flow of the U.S. volcano observatories into a National Volcano Early Warning System (NVEWS). Both the seismic and geodetic data are available in real time through NEIC. An American Association for the Advancement of Science review of the USGS Volcano Hazards Program conducted in 2007 strongly endorsed the implementation of NVEWS to develop an integrated, national framework for real-time, systematic, and cost-effective volcanic hazard monitoring (AAAS, 2007).

---

[1] http://earthquake.usgs.gov/monitoring/anss/, http://earthquake.usgs.gov/regional/neic/, http://volcanoes.usgs.gov/.

TSUNAMI WARNING

The National Oceanic and Atmospheric Administration (NOAA) oversees the U.S. Tsunami Program[2] with its mission to provide a 24-hour detection and warning system. The NOAA National Weather Service operates two tsunami warning centers that continuously monitor seismological data provided by the USGS from domestic and international seismic stations to evaluate earthquakes that have the potential to generate tsunamis. The tsunami warning centers also disseminate tsunami information and warning bulletins to government authorities and the public. NOAA uses the earthquake location magnitude and a system of buoys and tidal gauges as input into predictive tsunami inundation models. The Deep-Ocean Assessment and Reporting of Tsunamis (DART) network was substantially expanded in 2008 from 6 to 39 buoys as a result of the Tsunami Warning and Education Act of 2006 (NRC, 2011b).

METEOROLOGICAL MONITORING AND FORECASTING

Accurate forecasting of extreme weather events critically relies on a number of land-based and space-based observation and monitoring networks and continuous data from them. The full restoration of important weather, climate, and environmental capabilities to two planned space missions (NPOESS and GOES-R), including measurement of ocean vector winds, all weather sea-surface temperatures, Earth's radiation budget, high-temporal- and high-vertical-resolution measurements of temperature and water vapor from geosynchronous orbit, have been identified as key needs (NRC, 2008). The future status of existing, operational polar orbiting observational systems is uncertain; such systems also were not designed to capture strong winds or high waves (weather extremes).

Detailed weather observations on local and regional levels are essential to a range of needs from forecasting tornadoes to making decisions that affect energy security, public health and safety, transportation, agriculture, and all of our economic interests. As technological capabilities have become increasingly affordable, businesses, state and local governments, and individual weather enthusiasts have set up observing systems throughout the United States. However, because there is no national network tying many of these systems together, data collection methods are inconsistent and public accessibility is limited. NRC (2009) identifies short-term and long-term goals for federal government sponsors and other public and private partners in establishing a coordinated nationwide "network of networks" of weather and climate observation.

---

[2] http://www.tsunami.noaa.gov/.

STREAMFLOW MONITORING AND FLOOD WARNING

Flood-stage warning in the United States is the responsibility of NOAA's National Weather Service[3] using sophisticated numerical models that incorporate real-time precipitation data as well as the real-time streamflow data from the USGS stream gauge network. The USGS stream gauge network provides a long-term record of river flow in addition to real-time data in support of flood monitoring. A 2007 report from the National Research Council recommended expanding the USGS monitoring activities on rivers and called for a plan for a 21st-century river monitoring system for data collection, transmission, and dissemination (NRC, 2007).

PUBLIC HEALTH WARNINGS

The Centers for Disease Control and Prevention (CDC)[4] is charged with monitoring disease incidence and prevalence. The CDC surveillance system is designed to coordinate with the nation's departments of health and with hospitals regarding reporting of any unusual patterns in infectious disease, and illness or deaths resulting from radioactive contamination, poisoning, or other sources. Research is needed to continue to improve this surveillance system and to design best practices in response when a problem is detected (e.g., NRC, 2011c).

---

[3] http://www.weather.gov/, http://waterdata.usgs.gov/nwis/rt
[4] http://www.cdc.gov/.

REFERENCES

AAAS (American Association for the Advancement of Science). 2007. Review of the United States Geological Survey Volcano Hazards Program. AAAS Research Competitiveness Program. Available at http://volcanoes.usgs.gov/publications/pdf/aaas2007.pdf.

NRC (National Research Council). 2007. *River Science at the U.S. Geological Survey.* Washington, DC: The National Academies Press.

NRC. 2008. *Ensuring the Climate Record from the NPOESS and GOES-R Spacecraft: Elements of a Strategy to Recover Measurement Capabilities Lost in Program Restructuring.* Washington, DC: The National Academies Press.

NRC. 2009. *Observing Weather and Climate from the Ground Up: A Nationwide Network of Networks.* Washington, DC: The National Academies Press.

NRC. 2011a. *National Earthquake Resilience: Research, Implementation, and Outreach.* Washington, DC: The National Academies Press.

NRC. 2011b. *Tsunami Warning and Preparedness: An Assessment of the U.S. Tsunami Program and the Nation's Preparedness Efforts.* Washington, DC: The National Academies Press.

NRC. 2011c. BioWatch and Public Health Surveillance: Evaluating Systems for the Early Detection of Biological Threats. Washington, D.C.: National Academies Press.